SARDAR SAROVAR PROJECT
ON
THE RIVER NARMADA

Impacts So Far and Ways Forward

Volume 3

About the Editors

R. Parthasarathy is a Professor, Faculty of Planning and Public Policy at CEPT University, Ahmedabad. He has about 25 years of teaching and research experience and has been a Visiting Scholar at the University of California, Berkeley; Professor and later Director of Gujarat Institute of Development Research, Ahmedabad. He has been a member of many government committees and has also been a consultant to DFID, ADB and other national and international organisations. He also works closely with NGOs. His research interests include, water management, large-scale irrigation systems, natural resources, infrastructure management and environmental economics. He has published 3 books and over 70 articles.

Ravindra H. Dholakia has been a Professor of Economics at IIM, Ahmedabad since 1985. He has about 34 years of experience in teaching Economics to different groups like students, executives, policy makers and senior government officials. He has practical experience in high-powered policy making and evaluation bodies in both state and central governments such as the Sixth Central Pay Commission in India (2006-8), State Public Finance Reforms Committee (1998-2001), Expert Committee to Restructure Public Sector Units (2004-7) and Committee to Restructure Public Debt (2003-5), appointed by the Government of Gujarat. Currently, he is a member of the High Level Expert Committee appointed by the Government of India to suggest measures for efficient management of public expenditure (2010-11). He has published 39 monographs, 17 books, and about 130 research papers and international cases on economic planning and policy reforms.

Shri Narendra Modi
Chief Minister, Gujarat

Foundation stone of the Sardar Sarovar Dam on the river Narmada was laid on 5th April 1961 hence this Project will also be completing its long journey of five decades as Gujarat State is celebrating its Golden Jubilee. Sardar Sarovar Project is not only an icon of water resources development but it is also an expression of the strong will of the people of this State. Against a portrayed perception of SSP as a controversial project, this compilation reveals its multi-faceted dimensions. Not only the dimensions and benefits of this project are awesome but it has also got many innovative features in planning, designing, implementation, financing and managing. It is one of the most studied water resources projects in the world and its implementation has been a long struggle against all odds, which could be possible mainly due to wide public support and participation from people of Gujarat. Successive Governments made positive contribution in this effort, and the will of the people of Gujarat has been unstinting, and therefore in a true sense SSP is Project of the People of Gujarat.

Last one decade has seen accelerated development in this Project as a result of which the dam height could be raised to the spillway crest level, two hydro power houses could be made operational, 458 kilometer long Narmada Main Canal could be completed and command area of more than 5 lakh hectare could be developed by constructing a canal network of 19,000 kilometer length. Increased and assured availability of Narmada water for irrigation has helped in achieving and sustaining an impressive agricultural growth rate of more than 9% in the State against the national average of about 3% during the same period.

SSP is no less than a *"living technological university"* and the need for proper documentation of its evolution and growth was increasingly felt. As one of the firsts, these three volumes have now documented history of many facets of this very important project and have captured the highlights of this unique endeavor, while also suggesting ways forward.

The articles written by pioneer engineers, involved activists, academicians and government officials present various aspects of the project and Editors too have done a commendable job in bringing the individual contributions to a comprehensive publication level. Articles in this volumes address several concerns raised by critics at various stages of the Project. In my opinion on this 'World Water Day' during the golden jubilee year of the State, there can not be a better tribute to the people of Gujarat who have wholeheartedly supported the Sardar Sarovar Project in this journey. Sardar Sarovar Narmada Nigam Limited also deserves compliments for having supported this invaluable commemorative effort that will help the future generations to understand the evolution and growth of Sardar Sarovar Project.

(Narendra Modi)

Preface

The Sardar Sarovar Project (SSP) on the River Narmada has a long and interesting, yet sometimes controversial history. Even before the formal proposal was made in 1959 for a dam at the Sardar Sarovar site at Navagam, there had been various attempts to harness the Narmada. While Independence solved the problems of political complexity associated with organisation among the princely states for construction of the dam, the reorganisation of states on linguistic basis in 1956 added a new dimension. When preliminary work on the dam began in 1961, the disagreements between the riparian states on the sharing of waters led to the project being suspended. In 1969 the Narmada Water Disputes Tribunal (NWDT) was set up to decide the inter-state allocation of water and the costs of the dams and other infrastructure needed to exploit the river. In 1979 the Tribunal Award was made and full-scale construction of Sardar Sarovar began in 1987.

In its long history, SSP has faced many hurdles. It was not until the year 2003-04 that many of the issues relating to construction of the dam were resolved. Meanwhile, the main canal was constructed and many minors and sub-minors are in the process of being built. Other important features of this project are the rehabilitation of the project affected persons and the planning of use of its waters. While many of the controversies have been well documented, many salient features associated with the project have not been. It is precisely the many dimensions of the setting up of this project that acted as the stimulus to document the evolution and realisation of the project during the Golden Jubilee year of the state of Gujarat. To this end, three main themes emerged, namely, (a) the History of Design,

Planning and Appraisal including the Environment; (b) the Resettlement and Rehabilitation Process during Implementation; and (c) the Future Outlook and Way Forward.

The idea of producing edited volumes on these themes was discussed in a half-day Workshop at CEPT University on August 6, 2010. The Workshop, titled 'Evolution and Growth of Sardar Sarovar: The Life Line of Gujarat was attended by Engineers, most of whom have superannuated, but who had been involved in the design and construction of the dam and the canals; academics, NGO activists, serving and retired officials of Sardar Sarovar Narmada Nigam Limited (SSNNL); practitioners and policy researchers. In many ways this was an exploratory workshop; some came with the notion of processes being important; some focused on the history, and yet others suggested that the many innovations during the project reflect the flexible and sensitive responses to the changing environments and challenges moving away from the conventional designs. Continued discussions with the participants and the editors evolved into a commitment to contribute to the volumes well-researched articles on the different themes. It was also resolved that this volume need not go into the many controversies but should focus on processes and innovative aspects of SSP.

Thus the important feature of these volumes is not only that most of its contributors were directly involved in the project at one time or the other, but also had intimate knowledge of the various interrelated issues. As discussed in the August Workshop, the contributed papers before being published were to go through varied and multiple levels of review. The draft papers were discussed at CEPT University in the Second Workshop held from October 29 to 31, 2010. During this Workshop special invited lectures by Professor Y.K. Alagh (former Union Minister of Planning and Programme Implementation, Science and Technology and Power, GoI, and, Vice-Chairman of the Narmada Planning Group); Dr. C.C. Patel (former Secretary, Ministry of Water Resources, GoI, and formerly CMD of SSNL); Mr. Jay Narayan Vyas, Minister of Health and Tourism, and formerly Minister of Narmada Development, GoG); and Mr. I.P. Gautam (Municipal Commissioner, Ahmedabad and formerly, Director Finance, SSNNL) were also arranged. In fact, we later persuaded Mr. Jay Narayan Vyas and Mr. I.P. Gautam to convert their lectures into papers for the volumes.

During the Workshop the peer discussions on the papers contributed immensely to further sharpening the focus. Some of the authors also presented the final version of their papers during the two days of January 3 and 4, 2011 in the final workshop held at CEPT University.

An effort of this magnitude would not have been possible but for the strong and committed response of the authors. The papers were subjected to a refereeing process and some of them were re-reviewed before acceptance as publishable material. For this publication, we also had an Advisory Board consisting of Mr. N.B. Desai, Mr. N. Ramaswami, Mr. G.L. Java, Mr. P.H. Rana, Dr. Anil Patel, Mr. Hasmukh Patel, Dr. Sanat Chavan, Dr. Tushaar Shah and Dr. M.B. Joshi. We are indeed thankful to the authors, the referees and the Advisory Committee Members for their invaluable support in bringing these volumes within the time constraints. Our special thanks go to Mr. Vivek Khadpekar for his commitment and involvement with the editors while carrying out a commendable and a very competent editing of the papers.

All the analysis, views and opinions contained in the papers are those of the respective authors and do not reflect the views of the organisations they belong. Similarly, while we have taken all possible care to check the facts and figures presented in the paper, the errors of omission and commission still rest with the authors of the papers.

We are very grateful to Mr. S. Jagadeesan, Mr. K. Srinivas and Mr. G.R. Aloria of SSNNL, Gujarat for generously sponsoring and participating in the various workshops during this year of Swarnim Gujarat celebration. The present volumes are the outcome of these workshops. Dr. M.B. Joshi, along with his colleagues at the SSNNL, Dr. K.D. Acharya and Dr. Vinod Rana, provided very crucial support at all stages during the work including providing data, documents, maps and many reference materials. Ms. Mahima Gupta, an irrigation engineer and a planner herself, provided able research support; Ms. Aparna, Mr. Harish Joshi and Ms. Soumini Raja were always willing to help us not only during the organisation of the three workshops but also during the various stages of bringing out these volumes.

At the CEPT University, Dr. R.N. Vakil was always very encouraging right from the time the plan to bring out these volumes was discussed. In fact he speeded up the setting up of the CEPT University Press and the Centre for Documentation and Publication at the University to facilitate the work. Ms. Nirmala Khadpekar, Associate Professor at CEPT University took charge of the publication of these volumes and provided able and unflinching support. Professors Utpal Sharma and Shivanand Swamy not only provided crucial support but, along with Professor Ashwani Kumar, also attended some of the workshops and gave valuable feedback. We are also thankful to the Faculty members for their encouragement. A work of this kind requires administrative and library support. The CEPT University administration, particularly the Registrar, Staff at Centre for Research, Development and Consultancy, the library staff and Jitubhai and Dhuljibhai provided this critical support. We deeply appreciate and thank all of them.

The management team of the Concept Publishing Company was extremely cooperative and efficient in bringing out these volumes in time.

Ahmedabad
March 2011

R. Parthasarathy
Ravindra H. Dholakia

Contents

Part 2: Planning for SSP

Part 3: Appraisal for SSP

Volume 2: History of Rehabilitation and Implementation

Part 4: Resettlement and Rehabilitation in SSP

Abbreviations

AIBP	Accelerated Irrigation Benefit Programme
AISLUS	All India Soil and Land User Survey
AMC	Ahmedabad Municipal Corporation
APMC	Agricultural Produce Marketing Cooperative
ARCH	Action Research in Community Health and Development
ARWSP	Accelerated Rural Water Supply Programme
ASS	Argillaceous Sand Stone
BARC	Bhabha Atomic Research Center
BCM	Billion Cubic Metres
BHEL	Bharat Heavy Electricals Limited
BIP	Broach Irrigation Project
BIS	Bureau of Indian Standards
BMC	Bombay Municipal Corporation
BOC	Board of Consultants
BOT	Build-Operate-Transfer
CA	Compensatory Afforestation
CAD	Catchment Area Development
CAPs	Canal-Affected-Persons
CARE	Credit Analysis and Research Ltd
CAT	Catchment Area Treatment
CBR	Costs-Benefit Ratio
CCA	Cultural Command Area
CDO	Central Designs Organisation, Gujarat
CEO	Chief Executive officer
CES	Consulting Engineering Services
CGWB	Central Ground Water Board
Ch.	Chainage
CHPH	Canal Head Power House
CI	Confidence Interval

CICFRI	Central Inland Capture Fisheries Institute
CMCE	Critical Maximum Credible Earthquake
CRISIL	Credit Rating and Information Services of India Ltd
CSS	Centre for Social Studies
CSWCR&TI	Central Soil and Water Conservation Research and Training Institute
Cumecs	Cubic Meter Per second
CVC	Constant Volume Canals
CW&PC	Central Water & Power Commission
CWC	Central Water Commission
CWINC	Central Waterways, Irrigation and Navigation Commission
CWPC	Central Water and Power Commission
CWPRS	Central Water and Power Research Station
D&R	Design and Research
DACNET	Directorate of Economics and Statistics
DAP	Diammonium Phosphate
DBE	Design Basis Earthquake
DCC	Data Collection Centre
DDB	Deep Discount Bonds
DDO	District Development Officer
DDRP	Dam Design Review Panel
DF	Design Floods
DMAs	District Metering Areas
DOC	Divisional Operation Centres
DoEF	Department of Environment and Forest
DSC	Development Support Centre
DSP	Dam Safety Panel
DST	Department of Science and Technology
EDF	Environment Defense Fund
EIA	Environmental Impact Assessment
EPC	Engineering, Procurement and Construction
EPGA	Effective Peak Ground Acceleration
ERR	Economic Rate of Return
ESG	Environment Subgroup
FAO	Food and Agriculture Organization
FCA	Forest Conservation Act
FCI	Food Corporation of India
FEM	Finite Element Material
FFDA	Fish Farmers Development Authority
FMG	Five Member Group

FRL	Full Reservoir Level
FSD	Full Supply Depth
FSL	Full Supply Level
GAD	General Administration Department
GCA	Gross Cropped Area
GCEL	Gujarat Communication &. Electrical Ltd
GCMMF	Gujarat Cooperative Milk Marketing Federation
GDP	Gross Domestic Product
GEB	Gujarat Electricity Board
GEER	Gujarat Ecological Education and Research Foundation
GERI	Gujarat Engineering Research Institute
GGRC	Gujarat Green Revolution Company
GIS	Geographical Information System
GNFC	Gujarat Narmada Fertilizer Company
GoG	Government of Gujarat
GoI	Government of India
GoM	Government of Maharashtra
GoMP	Government of Madhya Pradesh
GR	Government Resolution
GRA	Grievances Redressal Authority
GRC	Grievances Redressal Cell
GSDP	Gross State Domestic product
GSFC	Gujarat State Fertilizer Company
GSI	Geological Survey of India
GWIL	Gujarat Water Infrastructure Limited
GWRDC	Gujarat Water Resources Development Corporation
GWSSB	Gujarat Water Supply and Sewerage Board
Ha	Hectare
HB	Hybrid
HH	Household
HP	Hand Pump
HYS	High Yield Strength
HYV	High-Yielding Variety
IAS	Indian Administrative Services
IBM-IFSSR	Inter-State Board for Management of Integrated Fisheries in the Sardar Sarovar Reservoir
IBPT	Irrigation By Pass Tunnel
ICAR	Indian Council of Agricultural Research
ICICI	Industrial Credit and Investment Corporation of India
ICID	International Congress on Irrigation and Drainage

ICOLD	International Commission on Large Dams
ICOR	Incremental Capital-Output Ratio
ICS	Indian Civil Services
ICSSR	Indian Council of Social Science Research
ID	Irrigation Department
IDA	International Development Association
IDBI	Industrial Development Bank of India
IFPRI	International Food Policy Research Institute
IIM	Indian Institute of Management
IITM	Indian Institute of Tropical Meteorology
ILO	International Labour Organization
IMD	India Meteorological Department
IMT	Irrigation Management Transfer
INR	Indian Rupee
IPCC	Intergovernmental Panel on climate Change
IR	Independent Review
IRM	Independent Review Mission
IRMA	Institute of Rural Management Ananad
IRR	Internal Rate of Return
IS	Indira Sagar
ISP	Indira Sagar Project
ISRO	Indian Space Research Organization
IWMI	International Water Management Institute
J&K	Jammu and Kashmir
JNNURM	Jawaharlal Nehru National Urban Renewal Mission
JP	Jayaprakash Narayan
KBC	Kachchh Branch Canal
LA	Loan Agreement
LF	Load Factor
LIC	Life Insurance Corporation
LPC	Land Purchase Committee
LPCD	Litres Per Capita per Day
LVFO	Lake Victoria Fisheries Organization
M&I	Municipal and Industrial
MAF	Million Acre-Feet
MC	Municipal Corporation
MCC	Main Control Centre
MCE	Maximum Credible Earthquake
MCM	Million Cubic Meter
MDDL	Mean Draw Down Level

MDE	Maximum Design Earthquake
MEQ	Micro Earth Quake
MHa	Million Hectares
MHM	Million Hectare Meter
MHUPA	Ministry of Housing & Urban Poverty Alleviation
MIS	Management Information System
MLD	Million Litres per Day
MNP	Minimum Needs Programme
MOEF	Ministry of Environment and Forests
MOUD	Ministry of Urban Development
MOWR	Ministry of Water Resources
MP	Madhya Pradesh
MPISSR	Madhya Pradesh Institute of Social science Research
MSP	Minimum Support Price
MSU	Maharaja Sayajirao University
MW	Megawatt
MWL	Maximum Water Level
NABARD	National Bank for Agriculture and Rural Development
NBA	Narmada Bachao Andolan
NCA	Narmada Control Authority
NCSDP	National Committee on Seismic Design Parameters
NDDB	National Dairy Development Board
NDS	Narmada Dharangrast Samiti
NEERI	National Environmental Engineering Research Institute
NFHS	National Family Health Survey
NGNS	Narmada Ghati Navnirman Samiti
NGO	Non-Government Organisation
NGRI	National Geophysical Research Institute
NHPC	Narmada High Power Committee
NIFDA	Narmada Inter-State Fisheries Development Authority
NIH	National Institute of Hydrology
NMC	Narmada Main Canal
NOF	Non-Overflow
NPDDC	Narmada Project Dam Designs Circle
NPG	Narmada Planning Group
NPK	Nitrogen, Phosphorus and Potassium
NPSSM	Narmada Project Steady State Model
NPUSM	Narmada Project Unsteady State Model (NPUSM)
NRSA	National Remote Sensing Agency
NRSC	National Remote Sensing Centre

NRW	Non-Revenue Water
NSA	Net sown area
NSP	Narmada Sagar Project
NSSO	National Sample Survey Organisation
NSV	No Source Village
NTPC	National Thermal Power Corporation
NVDA	Narmada Valley Development Authority
NWDT	Narmada Water Disputes Tribunal
NWDTA	Narmada Water Dispute Tribunal Award
NWRDC	Narmada Water Resources Development Committee
O&M	Operations and Maintenance
OBC	Other Backward Castes
OBE	Operating Basis Earthquake
OECD	Organisation for Economic Cooperation and Development
OECF	Overseas Economic Co-operation Fund
ONGC	Oil and Natural Gas Corporation
ORG	Operation Research group
PAFs	Project-Affected Families
PAPs	Project-Affected Persons
PAS	Performance Assessment System
PCR	Project Completion Report
PF	Provident Fund
PFC	Power Finance Corporation
PGA	Peak Ground Acceleration
PIM	Participatory Irrigation Management
PINS	Pressurised Irrigation Network System
PMF	Probable Maximum Floods
PMP	Probable Maximum Precipitation
PPP	Public Private Partnership
PUCL	People's Union for Civil Liberties
QL	Quintal
R & D	Research and Development
R&E	Research and Evaluation
R&R	Resettlement and Rehabilitation
RBPH	River Bed Power House
RCC	Reinforced Cement Concrete
RCNCA	Review Committee of NCA
RDSO	Railway Designs and Standards Organisation
RIDF	Rural Infrastructure Development Fund
RIS	Reservoir Induced Seismicity

RL	Reduced Level
RO	Reverse Osmosis
RRWHSs	Rooftop Rain Water Harvesting Systems
RTU	Remote Terminal Unit
RWSS	Regional Water Supply Scheme
SAR	Staff Appraisal Report
SBC	Saurashtra Branch Canal
SC	Scheduled castes
SCADA	Supervisory and Control Data Acquisition
SCBA	Social Cost-Benefit Analysis
SCF	Standard Conversion Factor
SD	Standard Deviation
SHGs	Self help Groups
SJMMSVY	Swarna Jayanti Mukhya Mantri Shaheri Vikas Yojana
SOI	Survey of India
SPF	Standard Projected Flood
SPIESR	Sardar Patel Institute of Economic & Social Research
SPPWCP	Sardar Patel Sahbhagi Jal Sanchay Yojana
SPV	Special Purpose Vehicle
SRR	Social Rate of Return
SS	Sardar Sarovar
SSCAC	Sardar Sarovar Construction Advisory Committee
SSD	Sardar Sarovar Dam
SSL	Suspended Sediment Load
SSNNL	Sardar Sarovar Narmada Nigam Limited
SSP	Sardar Sarovar Project
SSPA	Sardar Sarovar Punarvas Agency
SSRA	Sardar Sarovar Rehabilitation Agency
ST	Scheduled Tribes
STPR	Social Time Preference Rate
SWRC	Social Work and Research Centre
TCIL	Telecommunication Consultants India Limited
TDO	Taluka Development Offcier
TDS	Total Dissolved Solids
TECS	Tata Economic Consultancy Services
TISS	Tata Institute of Social Sciences
TMC	Thousand Million Cubic
UIDSSMT	Urban Infrastructure Development Scheme For Small and Medium Towns
UIP	Ultimate Irrigation Potential

ULBs	Urban Local Bodies
UN	United Nations
UNCED	United Nations Conference on Environment and Development
UNDP	United Nations Development Programme
UNICEF	United Nations Children's Fund
UNIDO	United Nations Industrial Development Organisation
UP	Uttar Pradesh
USBR	United States Bureau of Reclamation
USSR	Union of Soviet Socialist Republics
VLW	Village Level Worker
VSA	Village Service Area
WASMO	Water and Sanitation Management Organisation
WB	World Bank
WCD	World Commission on Dams
WHO	World Health Organization
WIHG	Wadia Institute of Himalayan Geology
WUA	Water Users Association
WWF	World Wildlife Fund

List of Contributors

Acharya, K. D., Assistant Engineer, Sardar Sarovar Narmada Nigam Limited, Gandhinagar. acharya_kishore2005@yahoo.co.in

Bassi, Nitin, Institute for Resource Analysis and Policy (IRAP), Hyderabad.

Chandra, A., Former Student, IRMA, Anand.

Chavan, Sanat, Former Principal Chief Conservator of Forests, Gujarat, A-2/3, Kabir Housing Complex, Opp. Don Bosco School Makarepura Road, Vadodara.

Das, Biswaroop, Professor, Centre for Social Studies, University Campus, Udhana Magadalla Road, Surat. dbiswaroop@gmail.com

Datta, K. Samar, Professor, Indian Institute of Management, Ahmedabad. sdutta@iimahd.ernet.in

Desai N.B., Former Secretary (Narmada), Government of Gujarat, 11, Triveni Park, Opp. S.N.D.T. College, Akota Stadium Area, Vadodara.

Desai, Vaidya Prerana, Sneh-Setu Foundation, Vadodara, snehsetufdn@gmail.com

Dholakia, H Ravindra, Professor, Indian Institute of Management, Ahmedabad. rdholkia@iimahd.ernet.in

Garg, Amit, Professor, Indian Institute of Management, Ahmedabad. amitgarg@iimahd.ernet.in

Gautam, I.P., Municipal Commissioner, Ahmedabad and Former Executive Director, Resources, and Director of Finance, SSNNL, ipgautam@egovamc.com

Goswami, Subhrangsu, Assistant Professor, Faculty of Planning and Public Policy, CEPT University, Ahmedabad. s.goswami@cept.ac.in

Gupta, R.L., President, Jaypee Ventures Limited, NOIDA, Uttar Pradesh

Iyengar, Sudarshan, Vice Chancellor, Gujarat Vidyapith, Ahmedabad. sudarshan54@gmail.com

Jagadeesan, S., IAS, Managing Director, Sardar Sarovar Narmada Nigam Limited, Gandhinagar. sjagadeesan53@gmail.com

Java, G.L., Former Director (SSNL) and Secretary (Retired), Narmada Water Resources and Water Supply Department, Government of Gujarat, gl7java@yahoo.com

Jhala, Neha, Research Associate, Indian Institute of Management, Ahmedabad, nehajhala@ iimahd.ernet.in

Joshi, M.B., Superintending Engineer, Sardar Sarovar Narmada Nigam Limited, Gandhinagar. dr.mbjoshi@gmail.com

Kankal, Bhushan, Research Associate, Indian Institute of Management, Ahmedabad, bhushank@ iimahd.ernet.in

Krishnan, Sunderrajan, Director, CAREWATER, INREM Foundation, Anand. carewater@gmail.com

Kumar, M. Dinesh, Executive Director, Institute for Resource Analysis and Policy (IRAP), Hyderabad. dinesh@irapindia.org

Mehta, Ambrish, ARCH Vadodara, E-702, Samrajya Complex, Near Fatehgunj Post Office, Fatehgunj, Vadodara. m_ambrish@hotmail.com

Mehta, Dinesh, Professor Emeritus, Faculty of Planning and Public Policy, CEPT University, Ahmedabad. dineshmehta@cept.ac.in

Mehta, Meera, Professor Emeritus, Faculty of Planning and Public Policy, CEPT University, Ahmedabad. meeramehta@cept.ac.in

Parmar, B.J., Former Secretary, Water Resources Department, Government of Gujarat, bjpsays@rediffmail.com

Parthasarathy, R., Professor, Faculty of Planning and Public Policy, CEPT University, Ahmedabad, r.parthasarathy@yahoo.com

Patel, Anil, Arch Mangrol, Rajpipla, District Narmada

Pathak, Mahesh, Former Chairman, Gujarat State Agriculture Commission and Former Executive Member, Narmada Planning Group, Government of Gujarat, mhpathaks@yahoo.in

Pathak, Swati, Research Associate, Indian Institute of Management, Ahmedabad, swatip@ iimahd.ernet.in

Pullabhotia, Hemant Kumar, Research Analyst, International Food Policy Research Institute, NASC Complex, CG Block, Dev Prakash Shastri Road (Opp. Todapur) Pusa, New Delhi 110012.

Rathod, K.G., Chief Engineer and Director(Canals), Sardar Sarovar Narmada Nigam Ltd, Government of Gujarat, Gandhinagar, dire-canal-ssnnl@gujarat.gov.in

Sah, D.C., Professor & Director, Madhya Pradesh Institute of Social Science Research, 6 Bharatpuri Institutional Area, Ujjain, Madhya Pradesh. dc_sah@yahoo.co.uk

Shah, Tushaar, Senior Fellow, International Water Management Institute, Anand. tushaar@irma.ac.in

Sheth, Pravin, Former Professor and Head of Political Science Department at Gujarat University, Ahmedabad. pravinsheth@hotmail.com

Srinivas, K., IAS, Joint Managing Director, Sardar Sarovar Narmada Nigam Limited, Gandhinagar. ksrinivas_ias@yahoo.co.in

Sudhir, C., Former Student, IRMA, Anand

Tomar, Shubhra S., ICSSR Doctoral Fellow, MPISSR, 6 Bharatpuri Institutional Area, Ujjain, Madhya Pradesh

Verma, Shilp, UNESCO-IHE Institute for Water Education, PO Box 3015, 2601 DA Delft, The Netherlands.

Vyas, Jay Narayan, Minister of Health and Family Welfare, Government of Gujarat and Ex-Chairman, Sardar Sarovar Narmada Nigam Limited, min-health@gujarat.gov.in

1

Sardar Sarovar Project: Background and Overview

R. Parthasarathy and *Ravindra H. Dholakia*

Background

The bifurcation of the erstwhile Bombay State into Maharashtra and Gujarat took place on May 1, 1960. Its objective was to provide separate identity and opportunity for self-determination in order to address developmental aspirations of the population of the two regions. Both the states took different paths and followed distinct development strategies and policies to achieve rapid economic and social development (see Dholakia, 1994). At the time of the bifurcation, Maharashtra's per capita income was about 10 per cent higher than Gujarat's (Dholakia, 1985). The share of primary sector activities in the total income was 46 per cent in Gujarat and 42 per cent in Maharashtra, making both the states largely agriculture-based economies (Dholakia, 1985). However, geographically a large part of Gujarat was drought prone and extremely water scarce. Surface irrigation opportunities and potentials were limited, making agriculture in the state rain-dependent and vastly vulnerable. Drinking water in thousands of villages and hundreds of urban centres was a serious problem. It led to consequent problems of health and hygiene resulting in higher mortality, particularly among infants and children. For its overall growth and development, Gujarat had to focus on the agriculture and animal husbandry sectors by tapping all possible sources and potential of irrigation in the state. The successfully launched co-operative movement among milk producers effectively addressed the problems of employment and poverty and ensured regular flow of income in the rural areas of the state. Problems of drinking water and irrigation, however, still remained.

The River Narmada is the largest perennial river in Gujarat, and a major irrigation dam project on it was already proposed and pending with the Government of India (GoI). The site of the proposed project was at Navagam near Rajpipla in Gujarat. The progress on this project was not satisfactory, and the feeling in Gujarat was that the state's aspirations were not being properly represented and communicated to the Centre. As a separate state with an independent identity, Gujarat could now directly represent itself in the Union. The newly elected Government of Gujarat (GoG) diligently took up the demand for speeding up the processes of getting approval from the Centre for the Navagam dam. The state regarded the project as inevitable and essential for its economic and social development. It was politically expedient to call it the "Lifeline of Gujarat", even though Gujarat would be only one of the four states benefiting from the project.

It became clear later that Madhya Pradesh (MP) and Maharashtra would benefit most in terms of the power generated by the project, getting respectively 52 per cent and 32 per cent, while Gujarat would get only 16 per cent. Similarly, Gujarat would have control over only about one-third of the water flow. The other three states—MP, Maharashtra and Rajasthan—would get to share two-thirds of the water benefits. Thus, the project though located in Gujarat is truly a national project. However, the leadership and *de facto* ownership of the project, and hence practical responsibility and accountability, has to a large extent been assumed by Gujarat.

It is interesting to note that the project is in its final stages now, however, it has been delivering partial benefits in terms of drinking water and irrigation since 2003-04, Gujarat's agriculture and overall economy have started showing its positive impact. During the first four decades from 1960-61 to 1999-2000, agriculture in Gujarat was a highly fluctuating and unstable activity. Annual growth rates of agricultural income during these decades showed wide variation, with the coefficient of variation ranging from 474 per cent during the 1960s to 641 per cent during the 1990s. This amply reflects the high degree of risk and uncertainty that the farmers in Gujarat had to face. After Narmada water became available, this variation in annual growth rates fell sharply during the last decade of 2000-01 to 2009-10 to about 210 per cent (Dholakia, 2010: 2303-09). The project is yet to realise its full potential. Currently not even one-third of the envisaged benefits are available in Gujarat. The total state income has shown remarkable acceleration in its average annual compound growth rate since the 1990s, when a major part of construction activities on the project was underway. It is expected that because of this project, Gujarat would continue to experience high agricultural and overall growth in the next two to three decades, contributing substantially to national economic growth and development (see Dholakia and Datta, 2010).

The following chapters in this volume—the first in a series of three—provide historical perspectives on this mammoth project, its design, planning and appraisal with a view to (a) understanding and appreciating different perspectives on this very controversial development project which attracted attention from the world over; (b) documenting some of the achievements of engineering and technological talents available within the country and the state; (c) recognising the world-class efforts put in by GoG in planning details of the project; and (d) providing opportunities to look at the mistakes committed, so that similar projects, even of smaller scale and dimensions can be better handled and implemented.

Overview

The present volume is organised in three parts: (A) history of the design of the project and related matters; (B) history of the planning process and related concerns; and (C) history of the socio-economic appraisal or the social cost-benefit analysis carried out for the project. Part A comprises five chapters, i.e. Chapters Two to Six; Part B has three chapters, from Seven to Nine and Part C has one, Chapter Ten.

In Chapter Two, Jai Narayan Vyas describes the Sardar Sarovar Project (SSP) as an engineering marvel. With its unique features and unparalleled dimensions, it occupies a remarkable place on the world map of water resources development. Apart from its manifold benefits in terms of long-term water supply, energy generation, food security and thereby sustainable development of Gujarat state, SSP has overcome many engineering and technological challenges during its journey from concept to realisation. It is one of the most studied water resources projects of the world and has successfully adopted path-breaking approaches to arrive at technological solutions at every step of its implementation. It is a forerunner in offering new dimensions and evolving fresh engineering standards be they in planning, design or implementation.

In the initial phase of its construction, the project engineers successfully surpassed unprecedented challenges. Some noteworthy achievements in design and execution include River Diversion Scheme, submersible cofferdams under flowing water conditions across deep river channel, excavation of open cut diversion channel and providing gated construction sluices. The treatment of the geological fault in the dam foundation required tremendous amount of concrete and steel and is, therefore, considered to be more challenging than the construction of any major dam.

The Narmada Main Canal (NMC) is the world's largest concrete lined canal with a discharge carrying capacity of 1133 cumecs at its head. The cross drainage structures on NMC bear testimony to the engineering and technological skills of the project engineers. From conceptualisation to planning, design, construction and operation, NMC also had faced many technical issues for which there were no readymade solutions. State-of-the-art techniques including mathematical and physical model studies, were used to arrive at safe and yet economical solutions.

SSP adopted many innovative and path-breaking approaches in planning, design, construction and operation, unparalleled anywhere in the world. In future too, the project is all set to incorporate further unique features like SCADA based Remote Monitoring and Control System, the longest Branch Canal—the 360 km long Kachchh Branch involving many technical challenges *en route*, passing through difficult topography and a desert area which falls in Seismic Zone V. Thus, the path had not been smooth, nor were there readily available answers. These make SSP an engineering marvel in its true sense.

Chapter Three, by G.L. Java describes the technicalities and innovations involved in the construction of SP and the detailed studies undertaken on various technical aspects. It describes how a good understanding of the potential of the Narmada basin and the utilisable quantity of water in it was obtained before the conceptualisation of the dam and the appurtenant works. Systematic gauge and discharge observations had begun since 1947 which helped several flood pattern studies in the command area by various research organisations, and in formulating flood control measures.

During the stage of conceptualisation, a number of officers were commissioned to look after the construction activities of the Main dam, Power house and the canal works. Many eminent international and national experts in various fields of design, construction and geosciences were engaged to supervise and guide the design and the field engineers. Importantly, the project team always aspired to make the project one of the finest examples globally. The chapter throws light on the salient features of the main dam, power house and main canal along with the layout of the dam and appurtenant works.

Extensive site level investigations were carried out to finalise an appropriate site for the main dam and to test the impacts of floods and energy dissipation arrangement. Also a number of scaled models were built. The model studies took almost a decade and half for eventual finalisation of the basic parameters, after the confirmation of which construction activities started.

A geological fault zone along the river bed presented a challenge during construction of the dam foundation. Seismic studies followed and it was found that any movement caused by an earthquake of magnitude up to 6.5 on the Richter scale would have no significant effect on the structure. Detailed studies and discussions led to a consensus that there was no evidence of its being a seismically active fault. It was however prudently assumed that the Narmada-Son lineament was seismically active and the project structures were accordingly designed. Treatment of the fault zone and other required treatments to prevent the sliding of layers were carried out to ensure adequate resistance through the foundation.

Instrumentation for the Sardar Sarovar Project was designed in consultation with experts and eminent engineers from India and abroad. Besides normal instrumentation for recording temperature changes, seepage, uplift, stress, strains and deflection, seismic instruments were installed to fulfil the critical need to monitor seismic activity and its impact within the body of the dam and in the reservoir vicinity. Both the SSP power houses—CHPH and RBUPH—were completed in all respects and are in regular operation. However, since the dam has not reached the designed height power generation is not at the designed potential.

Chapter Four, again by G.L. Java, discusses issues pertaining to seismicity in the context of SSP related both to earthquakes and their effects. It is pointed out that the Narmada Valley was known to be seismically active, but seismic studies reveal that no major stresses are being built up in the valley and the probability of major earthquakes (beyond Richter Magnitude 7.0) in this zone is very low.

The chapter discusses design of the features incorporated into the dam to surmount earthquake forces. Many renowned agencies were involved at various stages to consider a seismic design of the dam and the designs were scrutinised by an International Dam Design Review Panel. A Design Cell was created in CWC for vetting the dam design. It is clear that the project had full infrastructural capability to evolve a safe and sound design of the dam and other components. Despite all safety requirements being fulfilled in keeping with BIS and international standards, the *Narmada Bachao Andolan* (NBA) had expressed its fears about the adequacy of the dam design to withstand a severe earthquake and had filed a writ petition in the Supreme Court (SC) of India. It took six years for all the writs to be heard and debated in the SC. All the issues raised by the NBA were replied to and discussed in the affidavits submitted to the SC by the Chief Engineer, Dam, who was also the Technical Design Head. The second part of Java's paper contains many details relating to this aspect.

Chapter Five, by B.J. Parmar relates the final stage diversion of the river during construction of the dam. A detailed study of the topographical and geological features of the site revealed that it was not possible to divert the entire voluminous discharge through tunnels or by any other technique to allow construction. Hence, it was planned to divert the flow in stages, which were contemplated amid time constraints related to weather and to the quantity of stream flow in three stages. In fact, revisions were also made in the diversion schemes as the Government had advanced the construction programme by a year in order to provide water for irrigation and power generation. The completion of the final stage diversion was therefore planned for March-April 1993 in order to meet the various deadlines. These revisions demanded not only additional work but also precision as discussed in the chapter. The overall diversion works were a stupendous task involving high-precision and micro-level planning. An interesting account in the chapter shows how, at the end of June 1994, water was allowed to flow through the permanent sluices to safely join the river without submerging the construction work area.

SSP initially caused some apprehension about the high rate of silting of the reservoir and it was also feared that it would not serve its planned life. B.J. Parmar in Chapter Six narrates how this aspect was taken care of. The Narmada Control Authority (NCA)—the apex body looking after all the projects in Narmada basin brought factual information about the various studies conducted in this regard. These studies revealed that the apprehensions about silting of SSP were unfounded. The chapter shows that the expected useful life of the reservoir is likely to exceed the figure assessed, as a number of dams were already built upstream prior to the completion of SSP. These dams eventually serve to reduce the amount of silt and sediments reaching SSP. Parmar also discusses the catchment area treatments planned and implemented in the area directly draining into the reservoir. He demonstrates that these act as a stabilisation measure for erosion-prone areas and also reduce the rate of flow of silt from such lands into the reservoir.

The design aspects of the dam having been covered in the preceding chapters, Part II of this volume commences with a narration by Mahesh Pathak on the Planning aspects of SSP in Chapter 7. He describes how professional backup helped in the evolution of SSP. The chapter depicts the early investigations for the project and the role of various important bodies and groups formed by the Central and State governments. In order to resolve the dispute among stakeholders over the height of dam, GoI had appointed a High Level Committee of eminent engineers headed by Dr. A.N. Khosla on September 5, 1964. The noteworthy feature of the Khosla Committee was its approach to Narmada development planning. While Gujarat broadly endorsed the recommendations of the Committee, MP and Maharashtra raised strong objections especially regarding the height of the dam and about the control over distribution of Narmada waters. Consequently, GoI constituted the Narmada Water Disputes Tribunal (NWDT) to resolve the case. The NWDT award is of great historical significance and is regarded as a model award for its precise and well drafted sharing of project benefits and costs.

Another important body constituted by GoG on the World Bank's recommendation was a high level Narmada Planning Group (NPG) in order to develop a basic planning framework for SSP. In order to evolve an efficient and equitable water management planning framework, NPG initiated a number of studies and had detailed discussions with national and international professional experts. It also formulated an action plan on a priority basis to look after the health related issues pertaining to the dam vicinity area. The Tata Economic Consultancy Services (TECS) engaged in economic appraisal of SSP had reported that the project would yield much larger benefits than the costs involved with a benefit cost ratio of 1.84 and an IRR of 18.3 per cent, both evaluated at economic prices.

Several doubts raised by anti-dam activists had been proved wrong when drinking water supply emerged as the first major benefit of the project. Rajasthan had more or less abandoned all hopes of getting Narmada water and therefore, it was nothing less than a miracle when the waters actually arrived. The norms recommended by the NWDT for Resettlement and Rehabilitation (R&R) were perhaps the most liberal ever to be provided for an irrigation project in India. For speedy and effective R&R, GoG established the Sardar Sarovar Punarvas Agency (SSPA) in 1992. By July 2010, almost all project affected families had been resettled.

In the decade of eighties, SSP was criticised on environmental issues and was regarded as an environmental disaster. The chapter recalls the time when the World Bank, embarrassed by the unrelenting pressure of the anti-dam lobby, appointed an Independent Review Mission on Sardar Sarovar Project (SSP) in September 1991 under the chairmanship of Bradford Morse, and traces the events that followed. Mitigation of adverse environmental impact of the project was one of the major objectives of the environmental planning framework which was evolved on the basis of various EIA studies.

During its journey over five decades, SSP encountered many roadblocks but was eventually acknowledged as a model project. The major challenges it must now face are water management and command area development. If it succeeds in this acid test, the author notes, it can rightfully be proclaimed as a model project both in India and abroad.

In Chapter Eight, I.P. Gautam recounts the various challenges faced by GoG in financing SSP. The non-availability of WB assistance and OECF funding had led to considerable thinking and debate within the government and among sections of the society on alternative modes of funding. GoG had and the project authorities had limited options available to finance such a multipurpose mega-project, where construction of dam was at a primary stage, main canal work and power house construction had just started, and due to pressure from a few NGOs the active participation of partner states was also not obtainable. Under these circumstances GoG through its Special Purpose Vehicle—Sardar Sarovar Narmada Nigam Limited (SSNNL)—played a lead role in financing this project and also evolved unique mechanisms for the purpose.

Interestingly, the idea of public participation as a means of financing the project came up. The scheme of bond issue gradually came into picture and, because it was unique, gained the overwhelming response of more than 700,000 investors from all over the country. The successes of Deep Discount Bonds in public financing provided further encouragement for similar bonds later, but by way of Private Placements. Thus, the withdrawal of WB and OECF from financing the project was converted into an opportunity for public participation. As described in the paper, leading co-operative banks, PF Trusts, Charitable Trusts, Port Trusts, Insurance Companies, Banks and financial institutions were the major investors in the private placement of bonds.

Besides the wide range of institutions all over the country, the Cooperative Banks in Gujarat especially played a major role by participating overwhelmingly in the bond issues. Many of the members of these cooperative banks belonged to rural and semi-rural areas of the state and perhaps were able to directly appreciate the benefits of the project. However, short-term and mid-term loans for the construction of main canal, river bridges and cross drainage works were raised from nationalised banks during the stay on construction of the dam by the Supreme Court.

GoG had also borrowed a huge amount of money for the purchase of hydro turbines from three Japanese companies in spite of the fact that the state was to receive the smallest share of energy produced. At one point in time the non-completion of R&R work in the MP and

Maharashtra was blocking the progress of the project. At that juncture, GoG took another exemplary decision and financed the R&R in MP and Maharashtra. Billions of rupees of advance payment was made to both these states for the completion of their R&R.

In Chapter Nine, Sanat A. Chavan discusses the efforts to incorporate environmental safeguards at various stages of SSP. It talks about all the studies carried out on the environmental aspects and also provides statistics on forest land used for SSP and the efforts for compensatory forestation. The project was one of the first in India to accept the importance of EIA and incorporated it in the planning stage itself, even before the guidelines for such studies were prepared by MoEF-GoI. The chapter also states the number of studies carried out by government, NGOs and independent bodies making SSP one of the most exhaustively researched projects in India.

The environmental upgrading process aimed at converting Kevadia and the surrounding project impacted areas into an eco-tourism zone. Regarding compensatory afforestation in Kachchh the chapter states that it not only bore encouraging results but also provided employment and fodder to the local populace. A preliminary interpretation of the catchment area by GEER foundation, using satellite imagery, shows a distinct increase in the density-wise area in the catchment of SSP between 1985 and 1994. The chapter also discusses various aspects related to the Shoolpaneshwar Sanctuary area and the fact that many rare and endangered fauna and avifauna have been recorded there, out of which some had been considered extinct for over 60 years in the area. Of the various initiatives on the SSP area, this study highlights changes in the variety of fish and methods of fishing that happened, which in turn positively affected the livelihood of fishermen. Adequate measures were also taken to improve the habitats for sustainable fishing on the periphery of the dam reservoir. Mangrove plantations and studies on health aspects also point to the care the project took on the environmental front. Chavan states that the environmental improvement measures carried out by SSP were unparalleled anywhere in the country. However, the vulnerability of the ecological balance calls for strong measures and continuous environmental guidance to the user groups so that the ills of other similar projects are averted at the most opportune moment.

Besides the aspects of Design and Planning for SSP, this volume also presents a chapter on the appraisals done of SSP. In Chapter Ten, Ravindra Dholakia discusses the various cost benefit analyses done. He underscores the distinguishing features of the evaluation methodologies and considers the evidences on the economic rate of return (ERR) with sensitivity analysis. He states that the social cost-benefit analysis (SCBA) of a project is about judging its economic and social viability upfront by conducting economic appraisal on an *ex-ante* basis.

The economic appraisal of a mega-project such as SSP cannot be done on "mutually exclusive projects" basis, but only on "independent project" basis. Hence the decision method in the economic appraisal is of 'select-reject' type and not of 'ranking' type. Since SSP is a multipurpose, multi-state project; the appraisals have to be considered treating it in a national perspective. Compared to other projects of similar scale worldwide, SSP has one of the lowest proportions—only 2 per cent—of area inundated against the potential command area. Only four villages out of total 244 affected by the project are fully submerged. Its location is so ideal that in the Narmada river valley a huge amount of water can be stored with minimal effort. As the terminal project in the Narmada river valley, its main outputs are water and power, which have very high positive distributional influence in the drought prone districts.

Dholakia first outlines the conditions and the principles with which a careful identification of the items of costs and benefits should be done. Multiplier and growth effects should be considered for SSP as it had to be analysed under 'independent projects' framework. All costs and benefits must be measured in real terms or at constant prices and pecuniary costs and benefits should not be considered in SCBA. These costs should further be classified into direct tangible costs (construction, O&M, command area development, catchment area treatment costs); direct intangible costs (rare plant species, historical or archaeological monuments, village shrines, places of pilgrimage submerged in the reservoir that would be permanently lost); indirect tangible costs (abatement, compensatory afforestation); and indirect intangible costs (psychological costs of re-settlers). The benefits should also be classified as direct tangible (water and fisheries), direct intangible (parks, visuals—utilities for tourists), indirect tangible (increase in ground water levels, flow of tourists) and, indirect intangible (reduced seasonal and permanent migration of people/animals). According to him, none of the appraisals for SSP was comprehensive in considering all types of costs, and all possible benefits of the project.

The review of the appraisals of SSP highlights that the evaluation of costs and benefits was done on the project costs, costs of submerged land, catchment area treatment, environment and miscellaneous items, irrigation benefits (net irrigation benefits from cropping pattern, crop yields, and cost of cultivation and farm gate prices of crops), fuel wood and fishery benefits, power benefits, municipal and industrial water benefits along with some of the indirect benefits. Though all the SCBAs done by different agencies differ in the way they were attempted, a common feature was that each one took the most conservative estimate of the benefits of SSP and considered the costs most liberally. As a result of which the estimates of ERR and benefit-cost ratio of the project from the studies were on the conservative side. Each appraisal found the project highly viable—socially and economically. The chapter also briefly comments on the R&R issue, the package offered to the people in Gujarat, and the conditions in M.P. and Maharashtra.

PART 1

Design of Dam

2

Sardar Sarovar Project: The Engineering Marvel

Jay Narayan Vyas

> *"Destiny is not what you have got, but what you can make out of whatever you have got."*

In the history of the endeavour of humans to regulate and harness natural resources for their own welfare, the Sardar Sarovar Project (SSP) emerges as one of the few initiatives that have played a significant role in bettering the quality of life for millions. With its unique features and unparallel dimensions SSP undoubtedly occupies a remarkable place on the world map of water resources development. Apart from its manifold benefits in terms of long term water supply, energy generation, food security and thereby the sustainable development of Gujarat State, SSP has overcome many engineering and technological challenges during its journey from concept to realisation. In the truest sense it is a marvel or engineering. TIME magazine, in 1994 described the Narmada Project as one of the "Eight Modern Wonders *Abuilding*". Such recognition brings to the project the kind of glory that the Taj Mahal enjoys among ancient structures.

One of the most studied water resource projects of the world, SSP has successfully adopted path-breaking approaches to arrive at technological solutions at every step of its implementation. It has been a forerunner in terms of offering new dimensions and evolving fresh engineering standards be they in planning, design or implementation. The engineering issues dealt with in the different project components like dam, hydropower, canal, canal structures and so on, had no off-the-shelf-solutions available anywhere in the world. The old adage, 'Necessity is the mother of invention' – has been absolutely true in the case of SSP and has led to many distinctive innovations.

From the viewpoint of integrated development with a basin approach, SSP is part of a much larger plan to use the whole course of the Narmada – the fifth largest in the country – comprising 30 Major Dams, 135 Medium Dams and about 3,000 Minor Dams along the 1,312 km course of the river.

Sardar Sarovar is a concrete gravity dam across the Narmada, 1210 m long and with a maximum height of 163 m above the deepest foundation level. It is already built up to the crest level of the spillway at 121.92 m.

The Dam – From Alignment to Achievement

It will be the third highest concrete dam in India, the first two being Bhakra (226 m) in Himachal Pradesh and Lakhwar (192 m) in Uttar Pradesh. In the initial phase of its construction, the project engineers successfully overcame unprecedented challenges. Some of the noteworthy achievements in the design and execution include River Diversion Scheme, submersible cofferdams under flowing water conditions across deep river channel, excavation of open cut diversion channel and providing gated construction sluices. Ten temporary construction sluices, each of size 2.15 m × 2.75 m, were provided in the body of the spillway at RL 18 m. These were closed in February 1994. Another set of four permanent river sluices are provided at RL 53 m.

Construction of Cofferdam for diversion of river flow

Construction Sluices in the Main Dam

The treatment of a geological fault in the dam foundation involved 214,000 m^3 of excavation, 250,000 m^3 of rock excavation, 256,000 m^3 of pre-cooled concrete and 53,000 tonnes of reinforcement steel. Treatment of argillaceous sandstone in two layers on the right bank and red bole layer on the left bank was carried out by providing a grid of concrete shear keys (3.0 m wide and 3.5 m deep) parallel and perpendicular to the dam axis, with a total length of 12 km. With such vigorous treatment, the design of the dam allows for a horizontal seismic coefficient of 0.125*g* and covers the additional risk of reservoir induced seismicity. Thus the foundation treatment itself was more challenging than the construction of any major dam.

Treatment of Geological Fault in the Dam Foundation

In terms of the volume of concrete involved in gravity dams, this dam will rank the second largest in the world with an aggregate volume of 6.82 MCM. The first is Grand Coulee Dam in the USA with a total volume of 8.0 MCM.

For mass concreting, a vibratory Grizzly with two feed hoppers, two Grizzly feeders, two Two-way Gates, one Jaw-crusher and four conveyor belts inclined at a maximum angle of 15° running at a speed of 1.5-2.0 m/s were installed. The capacity of the conveyor belt system used was 1,000 tonnes/hr and it had a total length of 4.5 km.

The Aggregate Screening Plant had two reclamation tunnels, each with five electrically operated vibro-feeders of 500 tonnes per hour. Bulk cement was stored under pneumatic pressure in six cement silos with a total capacity of 12,000 tonnes. To take care of heat of hydration during mass concreting, a Chilling Plant comprising six chillers was used with total installed refrigeration capacity of 1,350 tonnes of refrigeration, and an Ice Plant comprising eight Ice Makers with an aggregate installed capacity of 300 tonnes per day of flaked ice. A fully automatic computerised Batching and Mixing Plant with a production capacity 330 m^3/hour was used.

The two cable cranes used for placing the concrete – each of 28 tonnes capacity and spanning more than 1.5 km – were the longest ever in the world.

Concrete placement Bucket with capacity 9 m^3, covering maximum width of 120 m

The spillway discharge capacity of the dam at 87,000 cumecs will be the third highest in the world after Gezhouba dam on Yangtze River (113,000 cumecs) in China and Tucuruí dam on Tocantins River (100,000 cumecs) in Brazil.

To negotiate the design flood radial gates, seven of 18.3 m × 18.3 m each for the chute spillway and 23 of 18.3 m × 16.8 m for the service spillway are to be provided.

The River Bed Power House (RBPH) with an installed capacity of 1200 MW is underground on the right bank of the river, about 165 m downstream of the dam. It has six Francis type reversible turbine generators, each of 200 MW installed capacity, supplied by Sumitomo Corporation, Japan and BHEL, India. These six units were commissioned in phases from February 2005 to June 2006.

The runner – with a diameter of 6.95 m weighing 120 tonnes, produced in one piece.

Lowering of Steel Ring in RBPH Access Tunnel and Machine Hall of RBPH

The CHPH is a surface power station in a saddle dam on the right bank of the reservoir, with a total installed capacity of 250 MW (5 × 50 MW). These five units have been commissioned in phases from August 2004 to December 2004. They can be operated with a minimum reservoir water level of 110.18 metres.

Narmada Main Canal

The Narmada Main Canal (NMC), with a length of 458 km in Gujarat, further extending 74 km into Rajasthan, is the world's largest concrete lined canal with a discharge carrying capacity of 1133 cumecs at its head. Designed to annually convey 11.7 BCM of water to quench the thirst of Gujarat and Rajasthan, it has been completed and water has been flowing all the way through since March 2008. It is proposed to use this canal not only for conveyance but also for storage of water to improve the response time of the system. At full supply depth, it can store 220 MCM – a volume that can serve domestic needs of a mega-city like Ahmedabad for the entire year. This storage, available round the clock, is more than the capacity of dams like Watrak and Bhadar in Gujarat.

Building the Lifeline – to convey the Elixir of Life

Mechanised Paver Lining on Side Slope for conveying 11.7 BCM annually

American Concrete Institute Award for Excellence, 1991: Canal Lining of Narmada Main Canal – Most Outstanding Concrete Structure in India

This unique contour canal has a command area of 1,845,000 hectares on its left bank. Along its length of 458 km up to the Gujarat-Rajasthan border, 634 structures have been built, of which 114 are control structures (head regulators, cross-regulators and escapes), 231 are cross-drainage works for crossing natural streams and other canals (aqueducts, siphon aqueducts, canal siphons, drainage siphons and super-passage) and 289 are communication structures (road and railway bridges).

Head Regulator of Main Canal – Upstream & Downstream View

The head regulator – off-take structure of the NMC at Ch. 0 km – is 82.6 m long with 5 radial gates of size 12.20 m × 13.50 m. A total 12,000 m^3 concrete was poured and 490 tonne steel was used in this structure.

Some of the structures on NMC are magnificent in terms of size, technical complexity, design and construction. Noteworthy structures like the aqueducts across the Orsang and Mahi, the siphons at the canals across the Watrak and Sabarmati, escape structure on the

Sabarmati, Y-Junction (off-take point of the Saurashtra Branch Canal), canal siphons across the Banas and Saraswati – all bear testimony to the engineering and technological skills of the project engineers. From conceptualisation to planning, design, construction and operation, the NMC has faced many technical issues. Each one deserves a special mention. The range of design challenges involved in these structures included high discharges leading to larger barrel size, typical foundation problems, huge loading, long-term consolidation settlement, design of leak-proof joints at locations vulnerable to differential settlement, seismic factors and dynamic response of the structure, magnitude of fall, longer spans, limitation of the available materials and products (for example bearings) and so on, for which there were no readymade solutions. State-of-the-art techniques, including mathematical and physical model studies, was used to arrive at safe and yet economical solutions. Research institutions of national and international repute were consulted to resolve many such issues. From a construction point of view, mechanised paver lining on the side slopes, dewatering on major river crossings, slip formwork for speedy construction etc. were firsts for India.

Orsang Aqueduct on NMC at Ch. 48.59 km

The 548 m long and 82.3 m wide aqueduct across the river Orsang is another unique structure with 27 spans of 20.30 m each. The superstructure, of three monoliths with three barrels each 7.0 m wide and 6.93 m high, rises 21.6 m above the river bed and carries a water flow of 1,126.38 cumecs. The foundation consists of piled piers and open abutments. This mammoth structure consumed 223,200 m^3 of concrete and 16,178 tonnes of reinforcement steel.

A large Aqueduct on river Mahi – during construction

Constructed for crossing the River Mahi at NMC Ch. 142.86 km, this aqueduct is 602.5 m long and contains 8 Barrels of Size 6.10 m × 7.6 m. A total of 387 thousand m^3 concrete was poured and 22,904 tonnes steel consumed for completion of the structure, which took more than 9 years.

Each of its 634 structures on the NMC is unique. To cite just one of them, at Ch. 196.188 km, the canal crosses a major river Watrak, which has a routed design flood discharge of 14,500 cumecs – about 16 times that of the NMC discharge at the crossing site (977 cumecs). Constructing a 410 m long canal siphon with 10 barrels of size 6.0 m × 7.25 m each, on a highly compressible clayey stratum (type CH), was a real challenge. After thorough investigations, alternative designs and interaction with experts, a three-fold remedial course of action was evolved, comprising sand drain as foundation treatment, staggered construction leaving gaps for differential settlement and pre-loading of the partly constructed structure. With a primary objective of pre-consolidation (90% in 125-166 days) and secondary objective of strengthening foundation strata, 2038 vertical sand drains, each of diameter 200 mm and depth 12 to 25 m (total length 42,653 m) were used.

For controlled preloading, in addition to the dead load of partly constructed barrels, superimposed loading of 23-26 t/m^2 was required. This was achieved with 600 cast in-situ concrete blocks and 6000 pre-cast concrete blocks (1.25 m × 1.25 m × 1.0 m each) with total 8800 m^3 of concrete. In addition to this, 130,000 m^3 of natural aggregates were used.

After achieving pre-consolidation in 125-166 days, the structure was unloaded and the concrete blocks were used in the protection works, launching apron and also in buoyancy concrete. Usable natural aggregates after unloading were employed in concrete work and other items of the structure.

Watrak Canal Siphon – Preloading with concrete blocks and natural aggregates

The Canal Siphon, across River Sabarmati at Ch. 229.92 km, is 518 m long and 71.25 m wide with a discharge carrying capacity of

881.634 cumecs. For constructing three monoliths with three barrels (box-structure) – each 6.25 m × 6.50 m height – supported over a raft foundation, 823,000 m^3 of earth-work and 213,750 m^3 of concrete were used. This siphon crosses the river at 15 m below its bed level.

There are 38 branch canals off-taking from NMC to serve a vast command of 1.8 million hectares in 15 districts. The largest of these is Saurashtra Branch Canal which is 104.46 km long and discharges 424.80 cumecs. To negotiate the peculiar land topography, with falling ground from Ch.0.0 to 59.0 km by about 52.0 m and thereafter, from 59.0 to 104.460 km, a rising ground by about 66.0 m before tailing into Bhogavo-II reservoir, three power stations at fall points and five pumping stations at lift points are planned. A total of 48 MW hydro-power is planned to be generated at the three fall structures at Ch. 5.15 km, 17.76 km and 32.94 m, each with a vertical drop of 10.9 m. Thereafter a total lift of 71 m is provided by a series of five Pumping Stations at Ch. 66.518 km, 77.165 km, 87.95 km, 93.205 km and 102.84 km, with pumping head ranging from 10.8 m to 17.025 m. In these five pumping stations, which are the largest in India, there are 26 Concrete Volute Pumps, each of 20 cumecs capacity, and 22 Vertical Turbine Pumps of 5 cumecs capacity each. A single Concrete Volute Pump of 20 cumecs capacity could serve the daily water requirement of Burj Dubai in just 48 seconds.

SBC Pumping Station No. 1 during construction

SBC Pumping Station No. 1 – after construction

The Canal Siphon across River Banas at Ch. 360.011 km of NMC is another exemplary case of engineering skill. The Narmada water flow of 387.415 cumecs will be carried across the river Banas by this 700 m long and 53.2 m wide Siphon structure comprising two monoliths of four barrels each, 5.50 m wide and 5.60 m. high, supported over a raft foundation. Deciding the exact location of this structure and its length was a challenge because of an undefined river

Physical Model Study for Banas Canal Siphon carried out at Irrigation Research Institute, Roorkee

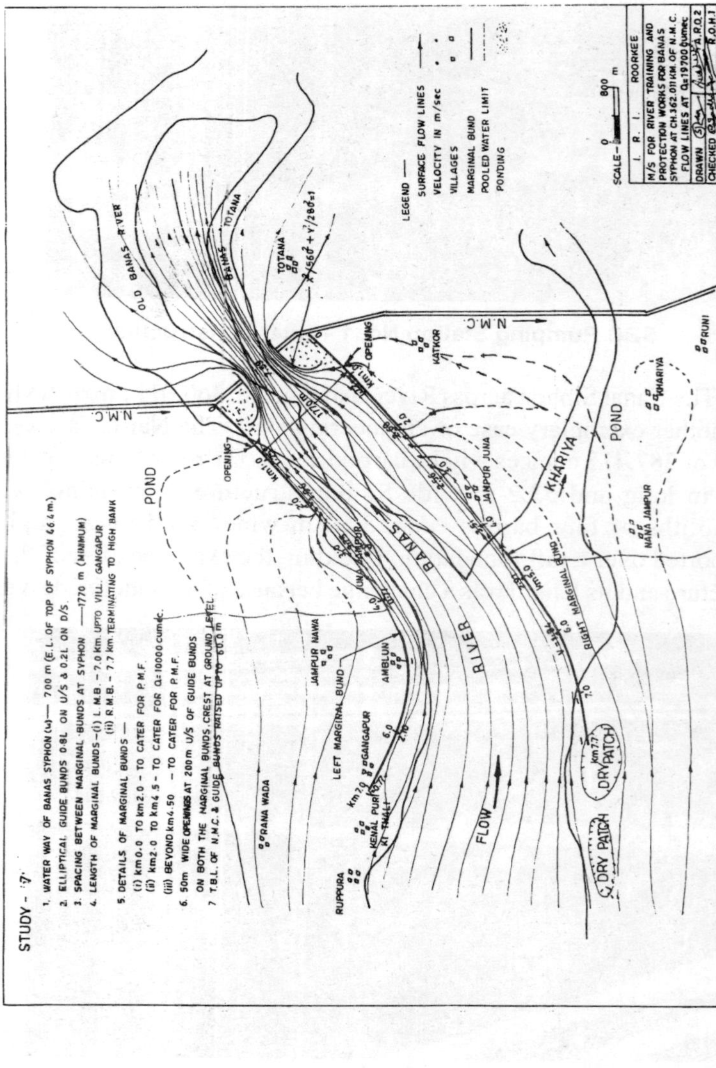

Model Study for deciding location and length of Canal Siphon across River Banas

gorge and peculiar hydrology causing flash floods in the form of sheet flow. Extensive physical model study at Roorkee and mathematical model study by CWPRS helped in arriving at the final decision. 203,761 m^3 of concrete and 13,453 tonnes reinforcement were consumed in making this mammoth structure.

As is evident from the above, SSP has adopted many innovative and path-breaking approaches in planning, design, construction and operation, none of which have a parallel in the world. In future too, the project is all set to have further unique features like SCADA based Remote Monitoring and Control System, the longest Branch Canal, the 360 km long Kachchh Branch involving many technical challenges *en route*, passing through difficult topography and a desert area which falls in Seismic Zone V. Thus, the path has not been smooth, nor were there readily available answers. But this only makes SSP an Engineering Marvel in its true sense.

Summing Up

This chapter is only an attempt to highlight some of the major features of Sardar Sarovar Project that immediately come to mind. My direct contact with the project ended when I relinquished office as the Minister in-charge of this mega-scheme. In preparing this chapter I have been considerably assisted and helped by my erstwhile colleagues, especially Dr. Mukesh Joshi. Hopefully this would place in some perspective the path-breaking efforts which blasted the myth that our engineers might fall short of the talent required in executing this unique project. The remarkable engineering features, some of them "First" and some of them "The largest" in the world, bear testimony to the technical skills, competence, talent and teamwork that our engineers and the contracting agencies put in to realise this dream. Equally, the sacrifices of those who were required to be resettled and the farmers and land owners who cooperated with the process deserve special mention. The tremendous emotional support that the NGOs and the people of Gujarat as well as people outside the state extended to the project is now a well known part of the history leading to this achievement. The contents of this paper may, therefore, be read as a minuscule description of many noteworthy achievements. I hope that they will be interpreted with these limitations in mind.

3

Sardar Sarovar Project: Features of the Main Dam

G.L. Java

TECHNICAL FEATURES OF THE MAIN DAM

Project Profile

The genesis of the Sardar Sarovar Project (SSP) lies in the dream of Sardar Vallabhbhai Patel and his close associate, Shri Bhailalbhai Patel, of the Narmada flowing to every corner of Gujarat. The plan to harness the river for irrigation and power generation in the Narmada basin commenced with a survey in 1946. Seven projects were identified. Four of these—Bharuch in Gujarat and Bargi, Tawa and Punasa in Madhya Pradesh (MP) were given priority for investigation. After the investigation, the dam at Gora in Gujarat was proposed, with a Full Reservoir Level (FRL) 161 ft. (49.80 m). Its foundation stone was laid on by the Late Prime Minister Pandit Jawaharlal Nehru on the 5th April, 1961. However, as more detailed and up-to-date contour sheets from the Survey of India became available, the possibility of raising the dam height for optimal water use was considered.

In 1964, to resolve the dispute about sharing of Narmada Waters between the Governments of Gujarat and MP, the Government of India (GoI) appointed an expert committee under the chairmanship of the late Dr. A.N. Khosla which recommended, in 1964, a higher dam with FRL 152.44 m (500 ft). However, no agreement could be arrived at and GoI constituted the Narmada Water Dispute Tribunal (NWDT) in 1969 under the Interstate River Water Disputes Act, 1956.

More than 60 years after Sardar Patel and Bhailalbhai dreamt their dream the Sardar Sarovar Dam on the Narmada is now complete after facing several hurdles along the way (described by other contributors to the present book). Located at 21° 50' N, 73° 45' E), the project is planned to provide irrigation to 1.8 million ha. in Gujarat, 75,000 ha in Rajasthan and 37,500 ha in Maharashtra. It will generate 1450 MW of power to be shared in the ratio of 57:27:16 between Madhya Pradesh, Maharashtra and Gujarat respectively. It will also provide drinking water to 8215 villages and 135 urban centres located in the drought prone region of Gujarat and two arid districts, Barmer and Jalore, of Rajasthan. Other benefits of the project include improvement in overall socio-economic environment and ecology in the western region of India.

The salient components of the project are:

(i) A main concrete gravity dam with a height of 125 m above the river bed, 163 m above the deepest foundation for creating a reservoir with gross and live storage capacities of 9500 and 5800 MCM respectively.

(ii) An underground River Bed Power House (RBPH) with six reversible turbines on the right bank, with an installed capacity of 1200 MW.

(iii) A surface Canal Head Power House (CHPH) with five conventional Kaplan type turbines having installed capacity of 250 MW.

(iv) A 460 km long main canal up to the Gujarat-Rajasthan border and 74 km long canal in Rajasthan with an extensive distribution system for irrigation.

Narmada River System

The Narmada, the longest west-flowing river of the peninsula, rises near Amarkantak in Anuppur District of MP, at an elevation of about 900 m, at 22° 40' N, 81° 45' E in the Maikala range. The river has a number of falls in its head reaches. At 8 km from its source, it drops 21-24 m at Kapildhara falls; and further about 4.6 m at the Dudhara falls. Its first major tributary, the Burhner joins the Narmada from the left, at the 248th km of its run. Following in a generally south-westerly direction in a narrow and deep valley, the river takes pin-head turns at places. At the 286th km from the source, it turns northwards and hardly a km further downstream it receives the Banjar, another major tributary from the left, and flows past Mandla town in a number of channels called Sahasradhara. Close to Jabalpur, 404 km from the source, the river drops nearly 15 m at the Dhuandhar falls, after which it flows through a gorge in the famous marble rocks at Bhedaghat.

Emerging from the marble rocks, the river enters the upper fertile plains and at the 464th km of its run, receives the Hiran, a major right bank tributary. Continuing to flow in a westerly direction through the upper plains, it receives further tributaries like the Sher, Shakkar, Dudhi, Tawa and Gangal from the left and the Tendoni, Barna and Kolar from the right.

Emerging from the gorge, the river enters the lower plains and meanders in broad curves until it reaches Bharuch. The Karjan from the left and the Orsang from the right are its important tributaries in this reach. Beyond Bharuch, the valley widens into an estuary discharging into the Gulf of Khambhat.

The total length of the river from the head to its outfall into the sea is 1,312 km. The first 1077 km are in Madhya Pradesh. The next 35 km form the boundary between MP and Maharashtra, and the subsequent 39 km, between Maharashtra and Gujarat. The last 161 km of flow is through Gujarat.

The Narmada Basin

The Narmada Basin covers an area of 98,796 km^2 between east 72° 32' to 81° 45' E and 21° 20' to 23° 45' N in the northern extremity of the Deccan Plateau. It covers large areas of MP and Gujarat and a comparatively small area in Maharashtra. The state-wise distribution of the drainage area is as under:

Drainage Area: State-wise

Name of State	*Drainage Area in km^2*
Madhya Pradesh	85,859 km^2
Maharashtra	1,538 km^2
Gujarat	11,399 km^2
Total	98,796 km^2

Source: Report of the Irrigation Commission, 1972.

The basin is bounded on the north by the Vindhyas, on the east by the Maikala range, on the south by the Satpuras and on the west by the Arabian Sea. It has an elongated shape with a maximum length of 953 km from east to west and a maximum width of 234 km from north to south. It has five well defined physiographic zones: (i) The upper hilly areas covering the districts of Anuppur, Shahdol, Mandla, Durg, Balaghat and Seoni; (ii) the upper plains covering the districts of Jabalpur, Narsimhapur, Sagar, Damoh, Chhindwara, Hoshangabad, Betul, Raisen and Sehor; (iii) the middle plains covering the districts of East Nimar, part of West Nimar, Dewas, Indore and Dhar; (iv) the lower hilly areas covering part of West Nimar, Jhabua, Dhulia and parts of Vadodara; and (v) the lower reaches covering mainly the districts of Bharuch and parts of Vadodara. The hilly regions are well forested. The upper, middle and lower plains are broad and fertile areas well suited for cultivation.

Utilisable Quantity of Water

It has been agreed by the party states and decided by the Tribunal (NWDT 1978) in its Order and Decision dated 8th October, 1974, that the utilisable quantity of water of 75 per cent dependability in the Narmada at Sardar Sarovar dam site should be assessed at 28 MAF. The distribution of this allocable quantum between the party states is as follows:

Madhya Pradesh	18.25 MAF
Gujarat	9.00 MAF
Rajasthan	0.50 MAF
Maharashtra	0.25 MAF

The Navagam Canal would carry 9.00 MAF for Gujarat and 0.50 MAF for Rajasthan. Distributing a 75 per cent dependable inflow of 27.01 MAF prorata between the party states, we get

Madhya Pradesh	17.61 MAF
Gujarat	8.68 MAF
Rajasthan	0.48 MAF
Maharashtra	0.24 MAF

Before the main dam and the appurtenant works can be conceptualised, it is necessary to understand the potential of the Narmada basin, its drainage area, river gradient, falls, utilisable quantity of water etc. and subsequently, for a multipurpose mega-project like the Sardar Sarovar, to understand its basic hydrology and arrive at the various likely floods that the SSP dam would be subjected to. These are Probable Maximum Floods (PMF), Design Floods (DF), Standard Projected Flood (SPF) and likely maximum water level (MWL).

HYDROLOGY

Discharge Observations and Run-off

Systematic gauge and discharge observations on the Narmada were started only in the year 1947. This is when the then Central Waterways, Irrigation and Navigation Commission (CWINC) took up investigations to formulate flood control measures and to assess water resources available for irrigation, hydro-power etc. Prior to that, only a few high flood levels appear to have been observed in connection with a bridge project and for flood protection works at Bharuch during the years 1907 and 1919.

The CWINC, later designated as the Central Water and Power Commission (CWPC) established 11 discharge observation stations on the Narmada during the years 1948-1951. Data from these reveals that at Mortakka, 59 km below the Narmada Sagar dam site, the catchment area entrapped is 67,190 km^2. Daily observations are available from 1948 onward, made with floats at a section 915 m downstream of a railway bridge on the Khandwa-Indore line of the Western Railway. At Garudeshwar, 11 km below Sardar Sarovar dam site, the catchment area entrapped is 89,345 km^2.

FLOODS

MPF and DF

Both Gujarat and Maharashtra have indicated figures for the 1000-year DF at Sardar Sarovar dam site. Gujarat has derived a figure of 0.87 lakh cumecs from a statistical analysis of the past records including the highest observed flood of 1970. Maharashtra has contended that the procedure adopted by Gujarat was not correct as certain values were neglected in the analysis. Moreover, the figures for the highest observed floods should be reduced on account of errors of observation. With these changes, Maharashtra derived a figure of 0.71 lakh cumecs. Gujarat has constructed a hydrograph for the 1000-year design flood of 0.87 lakh cumecs by proportionally increasing the ordinates of the hydrograph for the 1970 flood, taking the peak of that flood to be 0.7 lakh cumecs. If Maharashtra's contentions are accepted the peak of the 1000-year design flood would be only 0.71 lakh cumecs and the hydrograph for it would be the one prepared from the modified hydrograph of 1970 after making the necessary changes pointed out by Maharashtra.

The maximum observed flood (1970) at the Sardar Sarovar site and the consequently derived 1000-year flood there has two components—the inflow from the catchment above Narmada Sagar and that from the catchment below it. The former would get moderated upon construction of Narmada Sagar and upper dams, but not so the latter. In considering the MWL at Sardar Sarovar, the 1000-year design flood to be taken into account should be with moderated outflows from Narmada Sagar and inflows from the catchment area below Narmada Sagar.

Initial Studies for Probable Maximum Floods (PMF)

Since inception of the project in the year 1947 detailed studies have been carried out CWPC, New Delhi in 1952, India Meteorological Department (IMD), New Delhi in 1964 and Central Designs Organisation, Gujarat (CDO) in 1960 and 1974. These were mostly based on statistical methods. With the start of construction of SSP, studies have been further advanced by deterministic methods. The recent studies are by Narmada Project Dam Designs Circle (NPDDC), Gujarat in the year 1983, Central Water and Power Research Station (CWPRS), Pune in 1985 and National Institute of Hydrology (NIH), Roorkee in 1985.

For deterministic studies complete basin rainfall and runoff data had to be collected for the stations, most of them located in MP.

NPDDC collected these data from stations outside Gujarat through the following Central and State agencies:

(i) IMD, Nagpur;
(ii) IMD, Pune;
(iii) IMD, Ahmedabad;
(iv) Central Water Commission, Statistics Directorate, New Delhi;
(a) Hyderabad; (b) Surat; and (c) Bhopal;
(v) Field Offices of CWC;
(vi) Railway Designs and Standards Organisation (RDSO), Lucknow;
(vii) Madhya Pradesh, Director of Land Records, Gwalior; and
(viii) Maharashtra Agriculture and Irrigation Department.

Such data collected by NPDDC have also been furnished to CWPRS and NIH in order to facilitate their studies based on the same rainfall and runoff data.

Studies for PMF, as already indicated, were carried out initially by either single unit hydrograph, or probable maximum precipitation (PMP) approach, or stochastic method. Now flood studies are made by preparing sub-basinwise unit hydrographs and revised PMP approach. Stochastic methods comprised studies by flood frequency done by CWC, IMD and CDO prior to 1974. Supplementary data were obtained thereafter up to 1979. CWPRS and NIH also made similar stochastic studies before taking up studies by hydro-meteorological methods. The latter were done by NPDDC in accordance with the US Army Corps of Engineers, Hydrologic Model HEC-I. NIH studies were also done with the HEC-I model, while those by CWPRS were based on OPSET analogue model.

It may be noted that initial flood studies were done by flood frequency analysis based on observed gauge and discharge data of annual peak floods at Garudeshwar for the period 1948-1978. These data were supplemented with peak flows computed from observed gauge levels at Bharuch Railway Bridge, 100 km downstream of Sardar Sarovar dam, for 1887-1947. The Bharuch data were correlated with discharges at Garudeshwar. The studies made by the project in 1952 were reviewed by the Board of Consultants (BOC) in 1960.

IMD carried out studies in 1965 adopting the July 1944 storm and flood data of 1949, 1950, 1959, 1961 and 1962. Later, NPDDC, CWPRS and NIH did revised flood frequency studies with data for 1887-1980. They have also done studies using the Unit Hydrograph-Design Storm-Flood Routing approach.

In 1952 CWPC did flood studies using observed gauge data of three years from 1948 through 1950 at Garudeshwar, adopting one unit hydrograph common to the whole basin and a design storm of 100 years' frequency. PMF was estimated at 56,700 cumecs. This was reviewed by the BOC (appointed by CWPC) in 1960. They recommended increasing this peak by 10 per cent keeping in view that SSP would be the terminal reservoir and as such no chance could be taken. Accordingly the peak was revised to 62,300 cumecs. The IMD studies mentioned above determined a peak of 64,800 cumecs).

The emphasis of all studies done up to 1979 was on flood frequency analysis. CWC, while examining CDO's flood studies in 1979, pointed out the limitations of this method and emphasised that frequency studies cannot determine with reasonable degree of accuracy the volume and shape of PMF hydrograph and cannot evaluate effects of the upstream reservoirs correctly. These features are essential for routing floods through a reservoir and determining spillway outflow capacity. Consequently, emphasis was placed on hydro-meteorological studies. Frequency studies have also been revised by NPDDC and CWPRS. NPDDC did this task in continuous interaction with bodies like IMD, CWC, CWPRS and NIH.

Review of Flood Studies

The studies done by the various organisations were reviewed by Ray K. Linsley and J. Riedel, World Bank Consultants, and also by the Dam Design Review Panel (DDRP) and Dam Safety Panel (DSP). Riedel (1982) reviewed IMD's storm studies. Linsley analysed the flood studies by various organisations, including the PMF studies by NPDDC and CWPRS. He opined that the most extreme PMF could be expected to be of the order of 210,000 cumecs. This was arrived at assuming (a) a uniformly distributed maximisation factor of 1.40 instead of 1.25, thereby increasing PMF of 109,200 cumecs by 14 per cent; (b) a 5 per cent increase due to adoption of a more critical time pattern; and (c) a 50 per cent increase due to revised PMP.

DDRP opined that Linsley's PMP estimate of value was based on percentage increase on moisture maximisation factor, modified time distribution and the resulting PMP. Riedel indicated that the revised moisture maximisation factor could be higher or lower than 1.35. Further, he did not recommend a 50 per cent increase in PMP to be adopted for estimation of PMF.

Pursuant to the above recommendations of DDRP, analysis of the July 1927 storm, reportedly the heaviest over the subcontinent, was undertaken in consultation with Indian Institute of Tropical Meteorology (IITM), Pune. The data for this storm, which centred on Dakor in Mahi Basin, was transposed to the Narmada basin to compute PMP. The method of transposition adopted was slightly different from that used in the earlier studies. In those studies, it resulted in a peak of 166,500 cumecs. In the revised study a hypothetical combination of two storms—the 1926 storm which was intense in the upper basin and that of 1927—was transposed to the Narmada basin. The *in-situ* rainfall of 20-21 September 1926 and the transposed rainfall of 26-28 July 1927 placed back-to-back were taken to constitute the hypothetical 1926-27 storm to estimate PMP and PMF. Three-day isohyets of 26, 27 and 28 July, 1927 were transposed to the lower Narmada basin and the optimum transposition, producing maximum 3 days basin mean depth was determined. This transposition can be termed 'downstream-progressed' as compared to the earlier studies when the transposition was mainly 'catchment-centred'. The five-day design storm obtained with an arbitrary maximisation factor of 1.35 finally resulted in a PMF with a peak of 172,980 cumecs. PMF from various studies have also been routed through the reservoir initially at FRL 138.68 m.

From these studies, it was concluded that the PMF (inflow) in the Narmada at Sardar Sarovar would not exceed 171,100 cumecs. The corresponding outflow would be about 141,190 cumecs. Maximum reservoir water level would be RL 145.96 m and would remain for 45 hours above MWL of RL 140.21 m. There would be no overtopping of the dam.

Conceptualisation

The basic parameters of the Sardar Sarovar main dam, as has been stated elsewhere, were fixed during Tribunal Award. After the award was officially declared, the Government of Gujarat decided to set up an in-house design office at Vadodara. A full-fledged Narmada Project Dam Design Circle (NPDDC) office was created in the year 1978. Narmada Canals Design and Narmada Power House Design offices were also commissioned. All these offices were manned by a highly professional and dedicated group of young Engineers. This vital decision was taken to ensure that, besides continuity, the young engineers, who would be gaining in-house experience during their tenure, remained available for these works both in the design offices and in the field until the completion of the project. Similarly full-fledged field offices were also commissioned at Kevadia, to look after the construction activities of the Main dam, Power house and the Canal works.

In view of the World Bank Assistance, at the instance of the World Bank, Government of Gujarat created a Dam Design Review Panel (DDRP), consisting of eminent international and national experts in the various fields of design, construction and geosciences to supervise, guide and direct the design and the field engineers working on the project. Premier national and international institutions were also consulted for their expert advice and guidance as and when needed.

The young engineers working in the design offices and the field were encouraged to from time to time to visit ongoing and completed Projects in India and abroad to expose them to problem solving methods and procedures. It was imperative to ensure that the works in both design and the field were of the highest standards, conforming to international as well as Bureau of Indian Standards the specifications and to good engineering practices. Our aspirations were that, on completion of the project, it should be one of the finest examples globally of concrete dam and appurtenant works.

The Tribunal Award envisaged that SSP would be completed in three Phases, ending respectively ten, thirty and forty-five years from the commencement of work. The Tribunal orders with regard to the equitable allocations are made subject to review at any time after period of 45 years from the date of the orders.

After the final orders of the tribunal the works had to be started in right earnest, most efficiently and at the earliest, so that they could be completed on schedule. With the commencement of functioning of the NPDDC, work was started on the detailed design and drawings in-house for the dam and appurtenant works. At no point were engineering principles to be compromised or sacrificed. Standards and quality aspects were followed scrupulously and diligently.

As all the basic parameters had been already finalised by the Tribunal, and subsequently designing was done in-house and duly approved by DDRP and other expert organisations, it now became incumbent on us to proceed further and finalise the layout for the dam and all the appurtenant works comprising the project.

SALIENT FEATURES

The Sardar Sarovar Project envisaged the following features as per the final layout.

Main Dam

1.	Length of main concrete gravity dam	: 1210 m
2.	Maximum height above deepest foundation level	: 163.0 m
3.	Catchment area of river above dam site	: 88,000 km^2
4.	Gross storage capacity 7.7 MAF	: 9497.07 MCM
	Live storage capacity 4.75 MAF	: 5859.80 MCM
5.	Length of reservoir	: 214.00 km
	Maximum width	: 16.10 km
	Average width	: 1.77 km
6.	Spillway gates:	
	Service spillway 23 nos. (60' × 55')	: 18.300 m × 16.775 m
	Auxiliary spillway 7 nos. (60' × 60')	: 18.300 m × 18.3 m

Power Generation

1.	River bed power house (Underground)	: 1200 MW (6 × 200 MW)
2.	Canal head power house (Surface)	: 250 MW (5 × 50 MW)

Main Canal

1.	Length up to Gujarat-Rajasthan border	:	458 km
	Length in Rajasthan	:	74 km
2.	Base width in head reach	:	73.01 m
3.	Full supply depth (FSD) in head reach	:	7.60 m
4.	Design discharge capacity		
	(a) In head reach	:	1132 cumecs
	(b) At Gujarat-Rajasthan border	:	70.8 cumecs

Layout of the Main Dam and Appurtenant Works

The topography at the dam site required the head regulator of the Main Narmada Canal to be located almost 9 km from the main dam. The terrain of the right flank between the main dam and the head regulator is traversed by many rivulets and is very undulating. This fact was exploited to advantage by utilising the area between the head regulator and the dam to provide four balancing ponds.

Functioning of the Balancing Ponds

The ponds, created by building four rockfill dams on the rivulets, provide balancing storage capacity to regulate diurnal variations in discharge through the canal powerhouse (when operating as peaking station) at the fringe of Pond 1 and uniform withdrawals from the canal head regulator at fringe of the Pond 4. The requirement of balancing storage is assessed at about 1,308 ha-m. This capacity is available between EL 92.04 m and 95.09 m as stipulated by the Tribunal for operation of the pond. The water levels would thus fluctuate between EL 95.10 m and EL 92.05 m during normal operation. In addition to the driving head for flow of water from pond nos. 1 and 2 for regulating, a flood lift of about 2.44 m is provided for all ponds. Thus the anticipated highest water level in pond nos. 1 and 2 during floods would be about EL 99.06 m and that in pond nos. 3 and 4 would be about EL 97.53 m.

The range of operation of the ponds would therefore be small—about 3.05 m to 7.01 m compared to the heights of the dams—28 to 58 m. The upstream slopes of these dams would not be subjected to rapid draw-down conditions, and therefore not critical for their design. A common spillway for the four dams serves to dispose of floods from all the four streams. The spillway has been provided by cutting the hill beyond Rockfill Dam No. 3, incorporated a gated control structure and a chute along the hill slope terminating in a ski-jump, beyond which the long tail channel discharges water into the Panchmuli Nala downstream. The control structure with five gates (Godbole gates) will be operated automatically with variations in reservoir levels and have a discharge capacity of 595 cumecs. However at present, since lower discharges are required necessarily to be passed on to the river downstream, provision for mechanical operation to suit this requirement have also been made in the Godbole gates.

Rockfill Dams, Site Selection, Investigation and Geology

The topography, foundation conditions and seismicity of the region, and materials available for construction significantly influence the selection of type of dam and its design. For the rockfill dams built to create the balancing ponds, the sites have been selected after ensuring that the sections are optimum, commensurate with the length, height, geological and other related parameters. The section of the stream across which the Rockfill Dam-1 is built is relatively narrow. Overburden is shallow and good rock is available at shallow depth. For Rockfill Dam-2 the stream section is neither narrow nor very wide. Overburden is relatively thick ranging from 1 to 11 m. The rock is weathered to a considerable depth. For dams 3 and 4 the sections are wide, varying between 1430 and 514 m and the depth of the overburden varies between 3 to 17 m and 3 to 11 m respectively.

Samples of materials of overburden to depths relevant for stability analysis were collected at all the sites by taking trial pits along as well as upstream and downstream of the alignment. Undisturbed samples of foundation material were collected where feasible and tested to determine their field moisture content, field density, gradation, and shear parameters, i.e. angle of internal friction and cohesion. For stability analysis, 75 per cent dependable values of shear parameters and other properties, estimated separately for each rockfill dam, have been considered. The depth of the cut-off trench was based on the analysis of logs of drill-hole data, inspection of material obtained at different depths in trial pits and in the vertical faces exposed along the streams. While the Central Soils and Materials Research Station, New Delhi, had facilities for shear testing rock of maximum 80 mm size, it was, however, not equipped to test rock in the size range 150-600 mm, which was the requirement then.

Investigations were done and basic data collected at the four sites to help make an appropriate choice of design and construction for the dams. At Site 1, either masonry, earth or rockfill dam was feasible. At sites 2, 3 and 4 flexible types of dam were found to be more appropriate. Generally, rockfill construction is adopted at sites where foundation rock of reasonable good quality is available at shallow depth. Site 1 alone offered such a condition. The other three sites had sizeable thickness of overburden of a quality poorer than the rockfill materials used for the dams. Moreover, the depth of cut-off trench for Dam-2 varied from 4 to 12.5 m, whereas for the rockfill dams 3 and 4 it varied between 7-17 m and 6-11 m respectively. Angle of internal friction of overburden materials ranged from 25° to 27° against 40° for the rockfill material. Hence, sites 2, 3 and 4 were not very suited for rockfill dams. Because of the thickness of overburden and its poor quality, they would require loading berms at the upstream and downstream toes to load and stabilise the foundation.

For connecting the four ponds created by dams, link channels were necessary. Obligatory excavation for these channels, mostly in rock, would make freely available huge quantities of rock, estimated at 9 MCM. Unless proper use could be made of such a quantity of excavated rock it would pose problems of disposal. Under the circumstances it was decided to use the rock from the link channels to build the four dams adopting earth core rockfill construction as the most appropriate choice.

Geologically, the dam seat and its surrounding area comprises sandstone, limestone and shale of Bagh beds belonging to Upper Cretaceous system and Deccan trap of the Cretaceous Eocene Age with intruding dolerite dykes. Except the dolerite dykes, which are extensively weathered, the rocky outcrops are fresh. Some stretches along the alignment, which include inter-bedded layers of sand and gravel, have alluvial cover ranging in thickness from 0.15 to 15.25 m. All sites except Rockfill Dam 1 have thick cover of overburden material. Rockfill Dam-3 is the longest of the four and has quite a few complex geological problems.

Layout of the Main Dam, Energy Dissipaters and Power House Locations

The layout of the dam depends on many factors, the most important being the actual foundation conditions at the selected site. Other determining factors include the requirements of the type of dam contemplated, the spillway capacity to be provided and the locations of the power house and other appurtenant works. The Sardar Sarovar Main Dam has been envisaged to cater to floods conforming to BIS specifications. These stipulate that all major dams shall be designed for a return period of floods corresponding to 1 in 1000 years and 1 in 10,000 years. The dam shall also be checked for the PMF, SPF and DF. The spillway along with its appropriate energy-dissipation arrangements shall be designed for 1 in 500 years and shall cater to the dominant flood, which in our case is a 1 in 100 years return Flood. While finalising the layout of the main dam all these aspects have been considered and catered for.

This may sound quite simple, but to accomplish and satisfy all the requirements of the BIS codes and other engineering technical requirements is a time consuming and arduous task. To finalise the length of both the main and the auxiliary spillways, a number of alternatives were considered. Initially an in-depth study was done of other similar large projects in India and abroad. The DDRP and other eminent experts were also consulted. It was decided that the main spillway shall cater to the floods corresponding to 1 in 100 year return period, i.e. 66,260 cumecs and the auxiliary spillway for an additional 19,822 cumecs, i.e. both the spillways catering to a flood of 1 in 1000 year return period, i.e. 84,950 cumecs. After finalising the parameters it was necessary to test if the main dam would function properly under the various combinations of extreme forces occurring at any time. It was obligatory not only to satisfy all the requirements but also to test the various floods and their impact on the main dam and the energy dissipation arrangement being formulated. These tests, on scaled two-dimensional (sectional) and three-dimensional (hydraulic) models with all the required parameters in-built, were done under the aegis of NPDDC at CWPRS, Pune and the Gujarat Engineering Research Institute (GERI)—Gotri Hydraulic Laboratory, Vadodara. They served to confirm the behaviour of the structure of the main dam, the main and auxiliary spillways including the energy dissipation arrangements, and the appurtenant works of the dam.

The required model studies were done systematically for almost a decade and a half, running and re-running the models until the basic parameters were finalised. In the course of these studies the behaviour of the dam under various flood conditions was minutely observed after assigning the requisite parameters. Any discrepancies noticed were set right. Model studies for other components of the dam, power house and the appurtenant works were studied for almost two decades. Only when it was certain that all the parameters were in order was the layout finalised.

The total length of the dam is 1.2 km, consisting of 64 monolithic blocks conforming with the gorge width available at the site. The spillway length provided is 690 m, comprising 30 monolithic blocks, and 510 m of non-overflow (NOF) segment comprising 34 monolithic blocks. On the right flank there are 12 monoliths, of which six are of Power dam and six of the NOF segment. On the left flank there are 22 monoliths of NOF blocks, followed by 30 blocks of spillway—seven of the auxiliary and 23 of the main. The spillways together are capable of catering to the flood levels stipulated by DDRP and other experts, mentioned above.

During detailed discussions, N.G.K. Murthy, Expert Member and Consultant, specifically advised that the size of the spillway selected be such that under no circumstances would a man-made flood be caused. He explained that in the initial stages of most major projects the designers and the hydrologists are all intimately connected with design and hydrology as they are actively involved in formulating the project. Later, when they become old, experienced hands, they either get transferred or they retire. Those who are to take charge later may not necessarily have the same kind of intimate knowledge or insight of the particular site. Man-made floods have often been caused due to wrong operation and opening of the service-spillway gates not commensurate with actual site situations. To avoid such an eventuality it was essential to provide an optimum number of gates and a very clear gate operation manual giving the complete Rule Curve. His advice was followed to the letter.

The spillway models at CWPRS, Pune and the Gotri Research Laboratory, Vadodara are still being maintained for conducting necessary further studies. Also in case of an emergency or any other event such as a freak and unanticipated flood condition, the model should be readily available to study and resolve such an exigency and to study its effect on the structure. Any untoward effect that has not been contemplated or previously considered can thus be mitigated and the structure made safe. With all these basic parameters finalised, the planned construction activities could start.

River Diversion Scheme

The monsoon in this part of India lasts four months—June to September. The post-monsoon period of eight months, October to May, is available for construction. The river diversion scheme (Scope Report 2, 1972) envisages diversion of the post-monsoon flow through an open cut channel excavated in the river bed to the left of the deep river channel that normally carries post-monsoon flow. Two submersible cofferdams, one upstream and the other downstream of the dam seat area, were built across the deep river channel and two submersible bridges across the diversion channel for use during post-monsoon season. Construction sluices at foot of two spillway blocks were built in the diversion channel portion.

Construction Programme

The agency for building the cofferdams was fixed in December 1978. The first pile was planned to be driven by 15th January, 1979. According to the conditions of the contract, the river was required to be diverted by 1st May, 1979 to facilitate final closure of the cofferdam latest by 14th May. Rubble dumping and colgrouting (colloidal grouting) operations were to proceed concurrently, which was possible due to formation of compartments within the cofferdams. The cofferdams were planned to be completed by the end of May, allowing enough margin before the onset of the monsoon.

The deadline for other diversion works, such as excavation of the channel and submersible bridges, was the end of April 1978. The first stage of construction of river sluices in the channel were proposed to be completed in eight working months from December 1978, which meant that the work would spill over to the next working season and would have to be done under flowing water condition.

A brief description of the design and construction of submersible cofferdams as part of the diversion scheme for the post-monsoon discharge of the Narmada is given here. The cofferdams have been constructed by driving two rows of hollow steel piles down to the rock under flowing water condition with colgrouted rubble mass in between the two rows. The cofferdams have successfully withstood monsoon seasons.

The distinctive features of the cofferdams are as follows:

(i) Their construction was taken up concurrently with other diversion works and completed in a single season, making it possible to dewater the pool between them and tackle the dam foundation in the very next season.
(ii) Closure of the cofferdams could be successfully done even though the river could not be diverted in time for it. This was managed by allowing the river to flow temporarily through pipes specially embedded for that purpose in the cofferdams.

After completing the all the required diversion works and other ancillary works in the subsequent seasons, the cofferdams were dismantled.

MAIN DAM FOUNDATIONS

Geology at the Damsite

The left flank comprises eight lava flows of Deccan traps. These are primarily tholeitic basalts ranging from massive, porphyritic, amygdaloidal and vesicular with flows illustrating their individual characteristics influenced by rate of cooling. The right flank comprises five lava flows of Deccan trap, underlain by sedimentary deposits. Owing to a river channel fault, these lava flows are not continuous with those outcropping on the left bank.

The discontinuity of lava flows and sedimentary deposits from one bank to the other and the occurrence of crushed calcined and weathered trap, as revealed by 30 exploratory drill holes—19 vertical and 11 inclined, aggregating to a total length of 3262 m, established a fault along fair weather river channel.

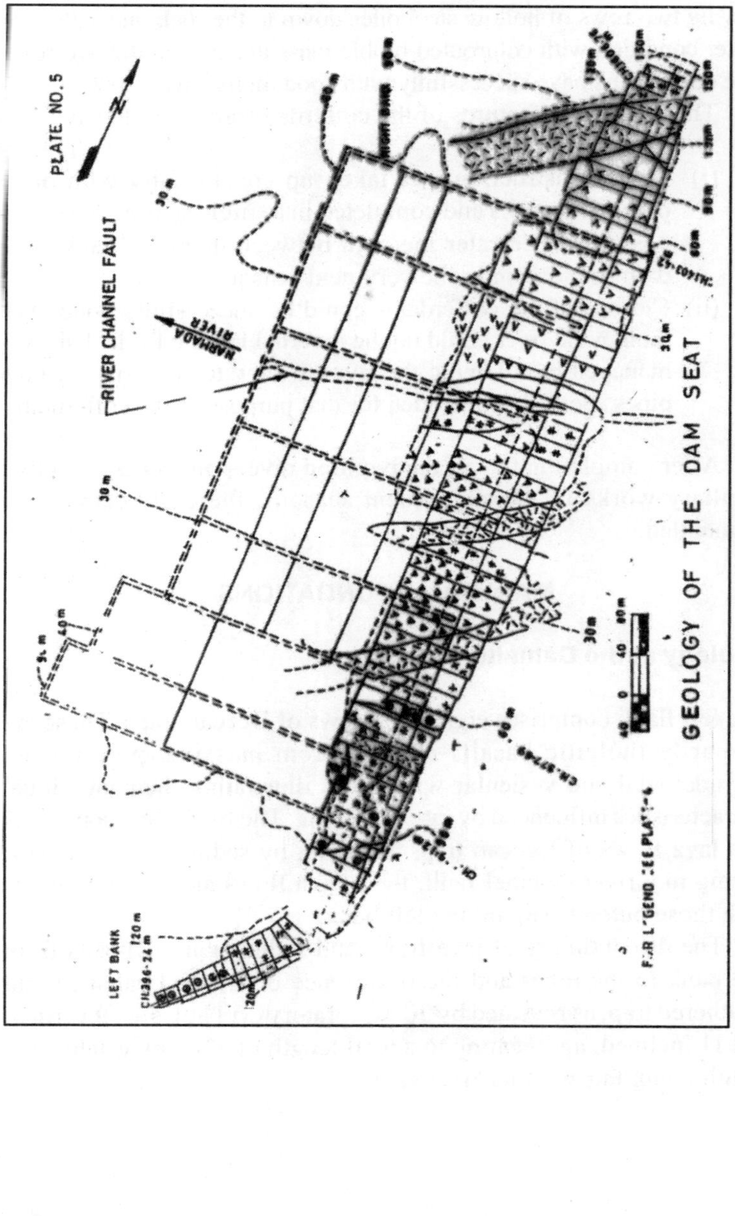

Geological Plan

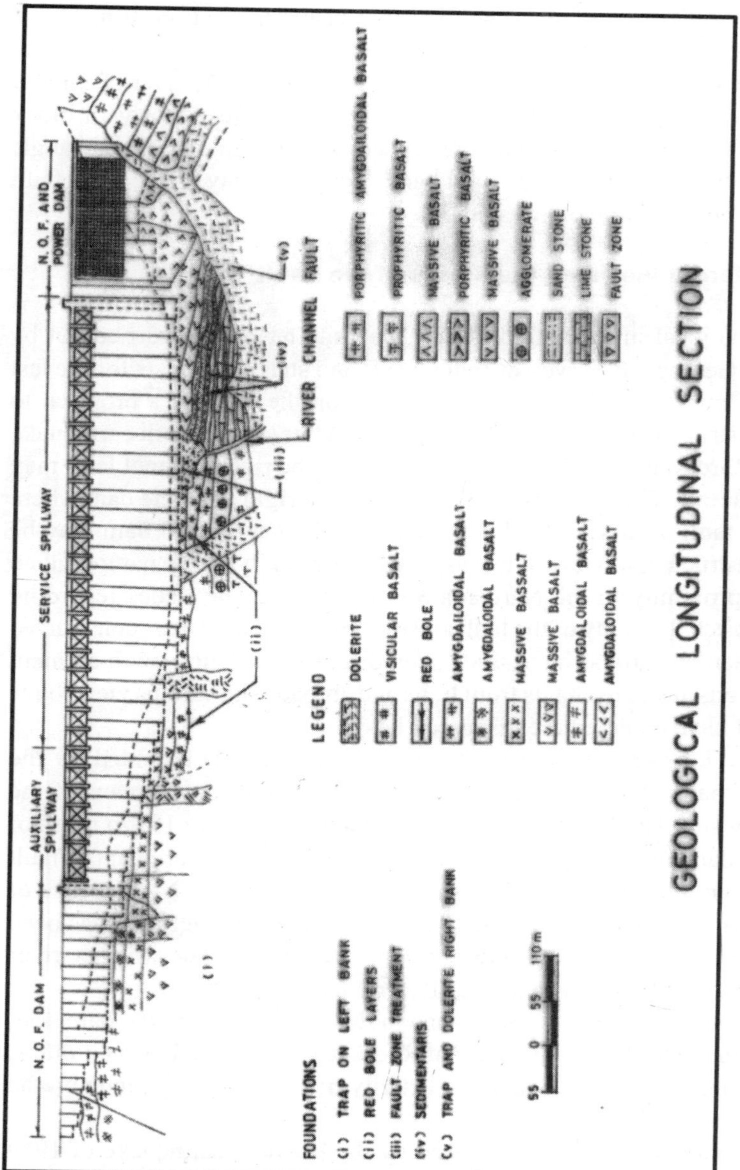

Geological Longitudinal Section

In 1996-97 a calyx hole (0.91 m diameter) was drilled to a depth of 36 m into the fault zone to enable visual inspection. It had been established earlier, based on drilled-hole data, that the fault zone width was 9.2-12.2 m. When the fault was exposed in the river channel it was found to be 5-7 m wide. It is a reverse type, foot wall block forming the left bank, with a throw down of 173 m. It dips at an angle of about 60° towards the right bank and has no clay gouge. This fault has been completely treated.

Different Views on the Status of the Fault

Geological investigations of the dam site have been carried out by the Geological Survey of India (GSI) and state geologists for the last several years. All investigations were on the basis of a proposal to construct a concrete dam at Navagam. After the Bharuch earthquake of 23rd March, 1970, GSI suspected that the river channel fault may be active. They suggested that instead of a rigid concrete dam where the fault occurs, a flexible structure such as a rockfill dam may be constructed, in view of its suspected seismically active nature, given its proximity to the Narmada-Son lineament. The issues regarding suspected activity of the fault and the type of the dam to be constructed at Sardar Sarovar have since been discussed at length with eminent experts and consultants from India and abroad. The various viewpoints that have emerged are discussed below.

The river channel fault, about 2 km long, is bounded by the Mokhadi fault upstream and the Akalbar fault downstream of the dam site. The Narmada-Son lineament located about 15 km north of the dam site is a block of *en echelon* faults. The river channel fault does not form a part of the Narmada-Son lineament. No evidence of recent movement along the fair weather river channel fault has been observed. A map showing the dam site and the fair weather river channel fault is given above (Refer: Geological Plan).

A micro-seismic survey in a radius of 15 km from the dam seat, carried out by the University of Roorkee over a period of six months, has indicated by and large an activity of sub-zero magnitude on the Richter scale.

The phenomena manifested by the Bharuch earthquake of 1970 have not provided conclusive evidence that the earthquake was due to activity along the Narmada-Son lineament. The epicentre of the earthquake is said to have been at the junction of the Narmada-Son lineament and the Cambay graben running north-south. Isoseismals prepared by GSI and iso-acceleration contours drawn by the University of Roorkee, however, are based on the assumed participation of the Narmada-Son lineament in the Bharuch earthquake.

The earthquake of 6.5 magnitude on the Richter scale (recommended for design of the Sardar Sarovar dam) would have a focal depth of about 15 km or more along a fault or lineament with a minimum length about 16 km. Since the river channel fault has a length of only 2 km, it seems unlikely that the focus of a 6.5 magnitude earthquake would lie in it. Hence, if at all an earthquake of this magnitude occurs in vicinity of the dam it could be only along the Narmada-Son lineament, about 15 km north of the dam site. Should any movement be caused by such an earthquake along the Narmada-Son lineament, its effect along the riverbed fault is expected to be insignificant.

The epicentre of the Koyna earthquake was very near close to the dam. Yet the Koyna earthquake did not cause any uneven settlement or relative movement in the dam foundation despite the presence of a 2-3 m wide shear zone in the seat of the dam.

The consensus reached after various discussions was that there was no evidence of riverbed fault being active. It might however be prudent to assume that the Narmada-Son lineament is seismically active and the project structures designed accordingly. While doing this it may also be noted that a seismic event would have two aspects—vibrations due to shock and physical relative movements along the fault plane. Design to withstand the vibrations was done for an agreed factor of 0.125 g. For the physical relative movements it was noted that the length of the fault would have to be sufficiently long to admit a sizeable movement.

Therefore, the possibility of significant movement along the fault plane within the seat of the dam is remote, as it peters out within 500 m upstream of the dam, well within its 2 km length.

With the need of a large spillway capacity, the site conditions at Sardar Sarovar are such that an overflow spillway is essential. This would preclude a rockfill dam from consideration. Apart rockfill dam on fault zone portion is not feasible as it is not possible to introduce a rockfill dam in a short length of about 60 m. Envelop junctions would need the spillway to be alternatively relocated in the hills of the left abutment. It has been worked out that the costs and quantities of work involved in a location of spillway in the hills on left flank are too prohibitive and hence not feasible. For the San Fernando earthquake of 1971 in the USA, accelerographs of Pacoima Concrete Arch dam recorded accelerograms equal to or slightly in excess of *g*. The abutment hills suffered heavily but not the dam. In the Koyna earthquake of December 1967 in India, the behaviour of the dam illustrated the competence of a concrete gravity dam to withstand a severe earthquake.

The DDRP for the Sardar Sarovar dam, comprising eminent engineers and geologists, observed that the riverbed fault was a local one occurring in the down-faulted segment, limited on either side by relatively younger faults. It showed no evidence of reactivation since formation. But its proximity to the Narmada fault had to be borne in mind while designing for the earthquake factor. The panel recommended that no further consideration need be given to the possibility of active seismic development at the base of the dam, and judged that the requirement of a large spillway to pass the estimated PMF of nearly 113,250 cumecs should determine the type of the dam. Given the extensive rocky strata in the surroundings, devoid of any alluvial deposit in the riverbed, the relatively narrow valley and the expectancy of very high floods during construction, a concrete gravity dam with an overflow section extending practically the entire flood riverbed was agreed upon as the basic layout of the work. The Koyna earthquake record adjusted to a peak acceleration of 0.25*g* was considered the maximum credible earthquake and the design earthquake was set at half this intensity.

Treatment of the Fault Zone

Diversion of the post-monsoon flow from the fair weather river channel was a must for the treatment of the fault zone. The diversion was taken up at the start of working season 1978-79 with a view to complete it in the same season, so that the treatment of the fault zone could start in the field season 1979-80. Due to the problem of shear zones in the foundations of the construction sluices, the diversion could not be completed on schedule, but the fair weather flow was diverted into the channel in May 1979. However, the treatment of the fault zone was commenced in the field season 1979-80. This was done by pumping out water from the pool in the fair weather river channel portion between two cofferdams, clearing silt and gravel from the dam seat portion, and taking up excavation of a trench in the fault zone. At the end of the field season the trench in the fault zone was excavated up to EL-7.0 m from the general foundation level of 18.0 m and bed level of about 12.0 m in the fair weather river channel.

Treatment of the fault zone consisted of excavation of the trench to remove the fault zone material, sheared rock adjoining the zone, some rock on the hanging wall side to required depth and back-filling it with mass concrete of suitable strength. CWPRS was requested to carry out photo-elastic studies for determining the depth of plug required for the treatment.

The concrete plug has been provided below five monoliths as the fault traverses in a skew direction from upstream to downstream. Excavation was done beginning from the upstream end going down to EL-16.5m and at the down-stream end to EL-6.5m. The plug is cast monolithically up to EL+18.0m, however suitable joints have been provided at the required intervals.

The design incorporates hammock reinforcement consisting of four layers of 36 mm diameter high yield strength deformed steel bars at 250 mm centre-to-centre spacing parallel to the dam axis in one direction, and parallel to the strike of the fault in the other. This reinforcement is to safeguard against any eventuality due to local weak pocket in foundation or heavy stress concentration in the concrete plug or unequal settlement which may travel upwards.

The hammock may also help in distributing loads uniformly in all directions below the concrete plug. The design of the concrete plug also provided a mat of reinforcement 36 mm diameter bars at 25 cm spacing both ways, in the final lift of concrete. This was for controlling possible shrinkage cracks and for preventing their propagation in the spillway blocks founded on the concrete plug. High yield strength (HYS) deformed steel anchor bars of 36 mm diameter, 8 m long and in a grid of 3 m × 3 m, were provided in the design for stitching jointed rocks and for enhancing shear resistance. This was achieved through an arrangement of split ends and wedges for anchoring them in foundation rocks.

An inspection-*cum*-drainage gallery of size 1.5 m × 2.3 m at EL 4.0 m in the concrete plug with its alignment parallel to strike direction

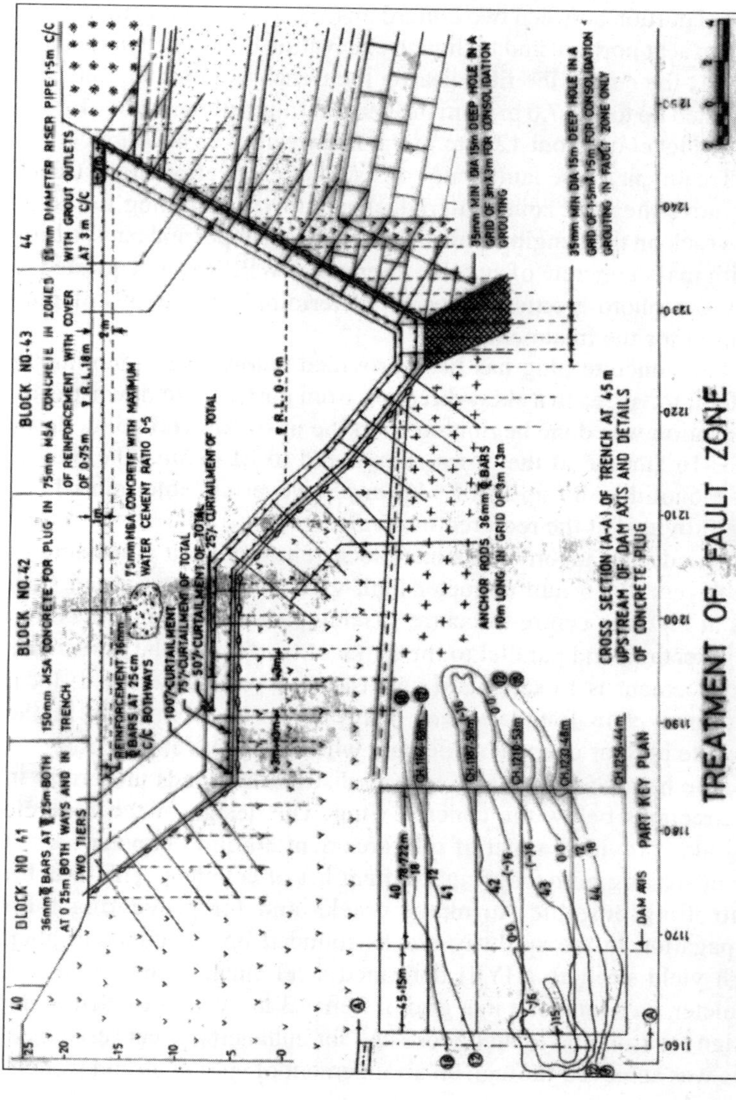

TREATMENT OF FAULT ZONE

of the fault zone was provided in the design. A row of black steel pipes of 100 mm diameter spaced of 0.75 m were provided along its alignment from bottom of the concrete plug, for future need to drill holes for grouting and drainage of the fault zone material. Two 2 m diameter vertical shafts, one provided with a spiral concrete staircase and the other open, were provided to the concrete plug for connecting this gallery with the foundation drainage gallery in the dam. A sump 30 m × 1.5 m × 2 m provided in the concrete plug at EL 2 m will be connected with foundation drainage gallery in the dam by a vertical shaft 1.5 m × 3 m for facilitating pumping of seepage water. As the inspection gallery in the concrete plug is provided deep inside foundations an elaborate arrangement of 1 m diameter vertical shafts at about 25 m spacing connected by similar horizontal circular opening is provided above its roof. This arrangement is further connected to two 1 m diameter vertical shafts in the concrete plug and the spillway block, to open into the atmosphere from the spillway bridge piers and ensure air circulation by natural draught. The inspection gallery would facilitate provision of instruments, drainage of foundation for controlling uplift pressures and continuous visual inspection for monitoring performance of the plug. Broad details of treatment of the fault zone are shown in the following photograph.

In the field season 1980-81 the trench in the fault zone was excavated up to EL-16 m upstream and up to EL-8 m downstream after carrying out obligatory pumping of water from stagnant pool in the fair weather river channel and clearance of silt deposited during 1979 monsoon. Quantities of excavation in overburden material and rock for the trench in the fault zone are about 214,000 m^3 and 250,000 m^3 respectively. The concreting of the fault zone was taken up in March 1981.

The concrete specified for backfilling the trench for construction of the concrete plug is pre-cooled to give placement temperature of 9°-12°C and compressive strength of 200 kg/cm^2 on 15 cm cubic specimens at an age of 28 days.

Season-wise Placement of Concrete in the Plug

Sr No.	*Working Season*	*Concrete (in cubic metres)*
1.	1980-81	7,200
2.	1981-82	73,000
3.	1982-83	113,000
4.	1983-84	62,800
	Total	256,000

All other details specified in the design have been implemented during construction. The treatment of fault zone involving 214,000 m^3 of overburden material excavation, 250,000 m^3 of rock excavation, 256,000 m^3 of pre-cooled concrete and 3,300 tonnes of reinforcement steel, cost about Rs. 175 million (at 1984-85 prices).

Treatment for Safety against Sliding of Argillaceous Sandstone Layers on the Right Bank and Red Bole Layer on the Left Bank

Extensive exploration of foundation identified weak layers [one red bole layer and two layers of argillaceous sandstone (ASS)] that required treatment to ensure adequate shear resistance through the foundation. Weak layers of ASS were encountered below the foundation rock on the right bank and one red bole layer was encountered on the left bank of the deep river channel. The stability calculations have specified criteria for design of solid gravity dams for a factor of safety against sliding. This necessitates provision of a suitable system of concrete shear keys both for treatment of ASS layers on right bank and red bole layer on left bank.

In layman's terms, the red bole layers and the ASS layers are very weak layers akin to slippery soft material between the hard rock layers, and may be the cause of sliding failure in the dam. The concrete shear keys that work as shear plugs safeguard to the dam against such failure.

Foundation treatment for ASS layers and red bole layer is fairly exhaustive. It covers Block Nos. 28–42 (395 m) on the left bank for treatment of red bole layer and Block Nos. 44-51, i.e. 184 m on the right bank for sedimentary layers. The treatment done is similar that provided at the Itaipu Project in Brazil. However, at SSP it is more extensive, covering the area by providing a grid of concrete shear keys parallel and perpendicular to the axis of dam along the ASS layers and red bole layer. Each shear key, longitudinal as well as transverse, is 3 m wide.

The treatment consists of a grid of a minimum of 10 shear keys parallel and two shear keys perpendicular to the axis of the dam. The width and the height (3 m × 4.5 m) of the shear key underground tunnel were decided in consultation with the field officers and geologists to permit easy movement of men and material. The average thickness of ASS was about 2.5 m for both layers, which are separated by a minimum of 3 m thick quartzite sandstone. The shear key is 4.5 m (i.e. 2.5 m + 2 m) in height. The bottom of the keys was placed just below the ASS.

For the two argillaceous layers at different levels, one of the issues was to decide the location of the upper layer grid *vis-à-vis* the lower layer grid, i.e. whether to locate the upper grid exactly vertically above the lower or to stagger them. By structural considerations and ease of construction the first option was preferable, as this would avoid concentration of stress on the rock mass within the grid along the lower level. It would also be a lot easier to drill for contact grout the lower level shear keys through the drifts (tunnels) along the upper layers. Hence, the work was carried using the first option.

The treatment of the red bole layer was also done on similar lines, except that in this case only one layer of shear key was provided as needed.

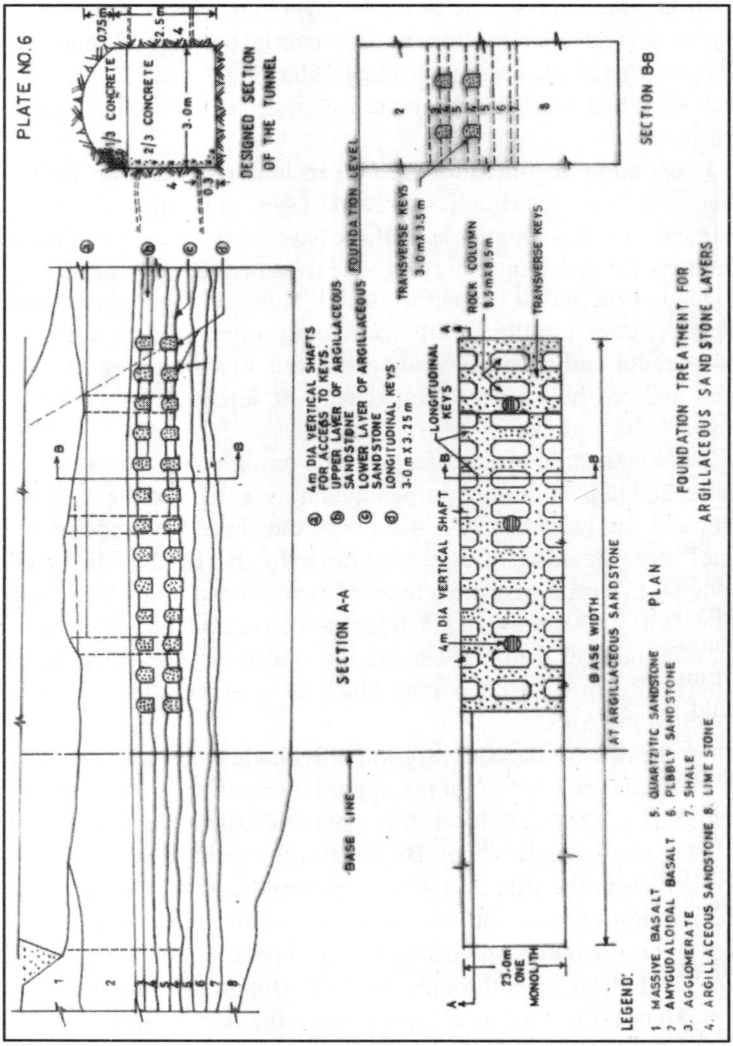

Argillaceous Sandstone Treatment

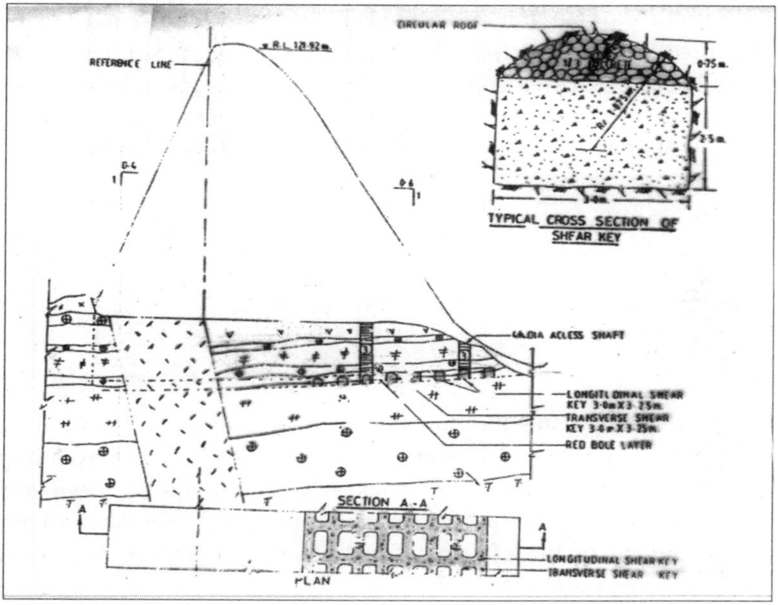

Red Bole Layer Treatment

For the shape of roof of the shear key, it had to be ensured that a good concrete rock contact was achieved, which is a prerequisite for safety against sliding. Initially the proposal contemplated pumped concrete with fairly high water-cement ratio and slump for the roof portion. However, during execution it was found that this method would not give the assured contact required between concrete and rock. Hence, concrete was resorted to and the complete treatment was provided by using concrete in two-thirds of the height of the keys and concrete in the remaining one-third height forming the crown. This method helped achieve satisfactory concrete-rock contact. The work for entire treatment was completed in the working season 1986-87 and the total cost was Rs. 150 million (at 1986-87 prices).

Spillway and Energy Dissipation Arrangements

The requirement of very large spillway capacity significantly influenced the final layout. The maximum observed flood at the dam site is 69,400 cumecs. Flood frequency studies carried out by

Government of Gujarat (1974) and submitted to NWDT have given a 1 in 1000 year frequency flood of 86,900 cumecs. Similar studies done by CWPRS yielded different frequencies as given below:

Estimates of Peak Floods for Different Return Periods

Sr. No.	*Return Period (Years)*	*Peak Flood (cumecs)*	*Peak Flood (million cusecs)*
1.	50	62,502	2.21
2.	100	68,251	2.34
3.	200	73,745	2.60
4.	1000	85,979	3.04
5.	10000	1,02,433	3.62

For safely passing the 1 in 1000 year flood with MWL 140.21 m, 27 spillway gates of size 18.3 m × 16.76 m are necessary. Considering a standby of 3 gates (10%), 30 gates were proposed, which included 23 for the main service spillway in the river bed and 7 for the auxiliary chute spillway on the left bank. The discharge capacity of 23 gates is 70,394 cumecs and that of 7 gates is 21,597 cumecs, making up a total discharge capacity of 91,991 cumecs. The design of the spillway is therefore adequate with a liberal margin to pass 1 in 1000 year flood.

The model studies by GERI and CWPRS, mentioned above, helped determine the most suitable type of energy dissipation arrangement to be provided for the service and auxiliary spillways. These studies, done over more than 7 years, examined the hydraulic efficiency, geological and foundation suitability and overall structural stability of the energy dissipater and the dam. Alternatives ranging from solid and slotted roller-buckets to ski-jump bucket and chutes have been studied at length. Based on their findings, the energy dissipation arrangement for the service spillway is a sloping-*cum*-horizontal jump type stilling basin. For the auxiliary spillway it is split-level chutes terminating in ski-jump buckets. During the estimates stage a 3 m thick slab was considered for the stilling basin, with HYS deformed anchor bars of 40 mm dia. spaced at 1.4 m c/c both ways in good foundation rock. For design of the stilling basin, statistics of actually observed floods were considered to arrive at the most dominant flood. The basin has been designed for 1:1000 year flood and has been checked for 1:10,000 year flood return period and the PMF.

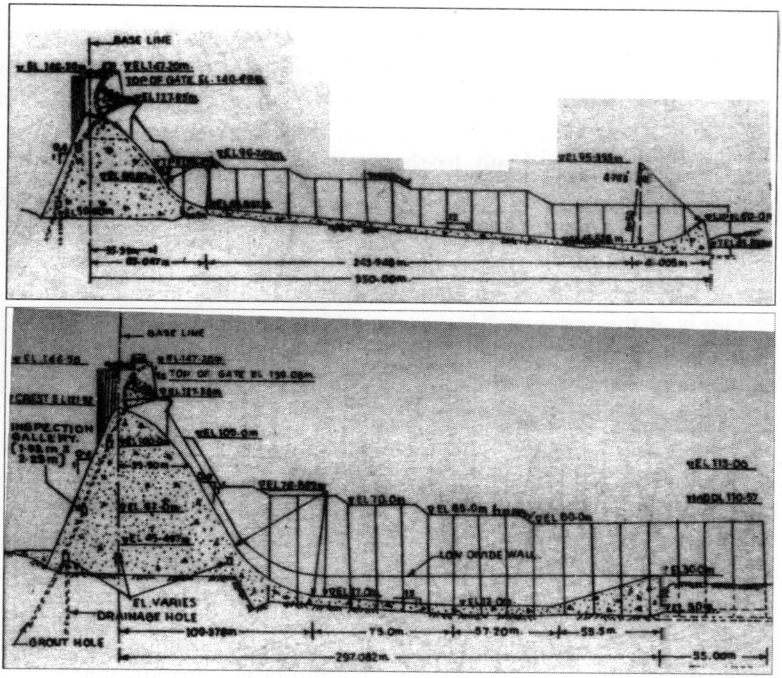

Longitudinal Sections of Service Spillway and Auxiliary Spillway

The structural design has taken into account the effects of Macroturbulence Theory based on the studies by Sanchez-Bribiesca and others presented in the proceedings of the 9th ICOLD (Bribiesca, 1967). Accordingly, the detailed structural design provides a 6 m thick slab at the upstream section (toe of the dam) tapering down to 3.5 m at downstream section, with 40 mm HYS deformed anchor bar, spaced at 3 m c/c. For protective measures downstream of the stilling basin and the chute, a 15 m long concrete slab covering the entire width of the service and auxiliary spillways was proposed. Below the energy dissipation, an elaborate drainage system has also been provided to prevent uplift and thereby failures. The drainage system below the stilling basin comprises a peripheral gallery, a central gallery parallel to the dam axis and a cross-connecting gallery from the downstream peripheral gallery to the central gallery. Underneath this is a grid of half-round drainage pipes with 7.5 cm dia. vertical drainage holes taken 10.0 m into the rock at each drain intersection. Along the half-round pipe drainage grid, spaced 11.5 m longitudinally and 15.0 m laterally, the water collected is conveyed through the drainage gallery to a sump behind the right bank training wall, from which it is pumped out to the river downstream. The sump is monitored and operated remotely with automatic pumps that are activated when water level in the sump reaches certain designed level. The sumps are located about 70 m at EL-4.0 m below right bank training wall. The total length of the drainage gallery below the stilling basin is of the order of 2 km.

Divide Walls and End Training Walls

For obtaining adequate conjugate depth the sill-level of the horizontal apron has been kept at EL 12 m. The entire service spillway of 23 spans has been further separated by low divide walls up to 25.0 m in order to safeguard the stilling basin from cross-currents and eddies in the event of non-operation of any two gates. The top level of the divide walls was determined on the basis of the GERI and CWPRS mentioned earlier, which analysed the effect of random closure of any two of the 23 gates and the resulting cross-currents and eddies observed in the stilling basin.

Keeping the sill-level at EL 12 m has resulted in a very high end training wall top with adequate free board keeping in view the maximum TWL likely to be reached for the 1:1000, 1:10000 and the PMF, which has been kept at EL 65 m. The design of end training walls and the divide walls between the stilling basins of the lower chute has had its own design problems. Detailed studies were carried out at CWPRS to study the hydrodynamic pressures on the low divide wall, the divide walls between the stilling basin and the chute, and also the end training walls. Special micro-pressure transducers were used in the model studies to determine the hydro-dynamic pressure on the walls and the floors.

Final designs for the low divide walls and the end training walls were done taking into account the effect of the hydrodynamic pressures. This exercise was informed by lessons learnt from Bhakra dam, which also incorporates a sloping-*cum*-horizontal jump type stilling basin. Its divide walls were designed with equal hydrostatic pressures acting from both sides so as to economise on the intermediate divide walls. Because of this one provision, when damages were noticed and had to be repaired in the floor of the stilling basin, the treatment not only became exhaustive and expensive, but pneumatic caissons had to be used to avoid dewatering of the pool.

SSP further innovated on the design of the divide and end training walls. Detailed hydrodynamic studies were done with CWPRS and the results incorporated in the design. Hydrodynamic and the hydrostatic pressures were taken to act on one side (for the condition where one side of the divide wall is filled and the other side empty). This resulted in providing heavy reinforcement to withstand the pressures in the RCC cantilever walls, which were 13 m in height. This provision would make the repairs/inspection and the maintenance work safer and simpler, avoiding the use of pneumatic caissons.

Providing Aeration Measures for Safety Against Cavitation Failure in the Spillway

The design of high head structures with high flow velocity requires prediction and evaluation of possible cavitations, cavitations failure of the proposed forms and the development of safety measures against cavitations failure. Full-scale cavitations investigations conducted in the USSR, USA and other countries have developed very simple methods against cavitations. The anti-cavitations protection method comprises measures whereby air is admitted to the cavitations zone. For the Sardar Sarovar dam adequate precautions were taken to ensure against cavitations damage to the service spillway, which spillway will be subjected to high velocity as well as high discharge concentration. Although all possible efforts are considered during design and contemplated during construction of the spillway, possibilities of cavitations erosion occurring at such high velocity cannot be ruled out. The principal cause of cavitations erosion on spillways usually is local irregularities in concrete surface, either due to imperfections in construction or improper and non-standard formwork, change of formwork not congruent to the design profiles and passing of sediments during service of the spillway. To safeguard and protect the Navagam spillway, aeration devices consisting of ramps, grooves and air supply ducts were proposed in the service and auxiliary spillways.

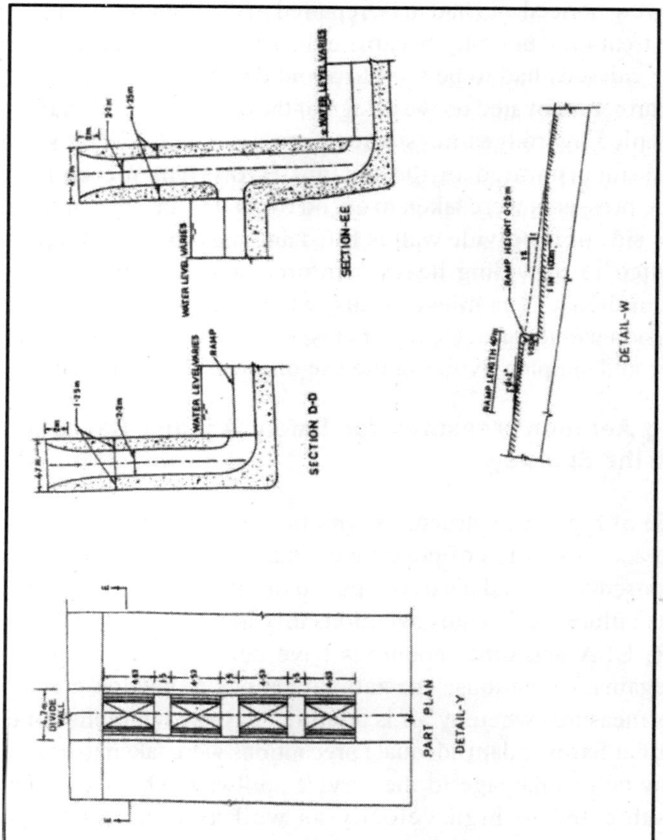

Aeration Measures for Safety against Cavitations Failure in the Spillway

Instrumentation

Instrumentation for the Sardar Sarovar Project was designed in consultation with experts of the DDRP comprising eminent engineers from India and abroad. For detailed layout of the instrumentation in the dam, IMD, CWPRS and other expert bodies were also consulted. The latest instruments, worth about Rs. 30 million, were acquired for the main dam and the seismic stations around the reservoir periphery.

The objectives of the instruments are twofold: Those embedded in or installed at the surface of dams and other hydraulic structures constantly monitor their performance in service and indicate the distress spots which call for remedial measures. Most hydraulic structures are built on conservative assumptions to provide for the unknowns in the design. Observations from the instrumentations will help to reduce these unknowns and place the future designs on a sounder footing.

Besides the normal instrumentation for recording temperature changes, seepage, uplift, stress, strains and deflection, there are seismic instruments installed to fulfil the critical need to monitor seismic activity and its impact within the body of the dam and in the reservoir vicinity. For SSP this scheme consists of nine seismological stations. Four each of these are on either side of the reservoir in MP and Maharashtra, and one is at Kevadia. These are meant to investigate the sources of seismic activity in the vicinity of Sardar Sarovar dam and also to evaluate any seismic activity that may be associated with the filling of the reservoir. These data are meant for use by earthquake engineers in the earthquake-resistant design of structures to be built in this region, and also may give important information to seismologists with regard to local earthquake mechanisms. In fact the last earthquakes which were felt near Bhavnagar in the Gulf of Khambhat in March 1989 on the 16th and at Chuda near Limbdi, south-west of Surendranagar on the 20th, were both recorded at the Kevadia observatory as 2.0 and 4.0 respectively on the Richter scale.

For proper recording and data, an online recording system with modern computers is installed both at the dam site and at Vadodara, to have on hand and to analyse the complete output data and results.

The information obtained through measurements promotes understanding of the influence of various parameters on structural behaviour. Surveillance of the behaviour of the structure and its foundations during operation is essential to guarantee the safety of any large dam. With the sophisticated instruments and computers installed on the Sardar Sarovar Dam and the ample precautions taken, there will be no difficulty in monitoring this large project, and in guaranteeing the safety of the dam.

Sloping Struts

In the service spillway, the foundation levels of blocks 41-47 and the fair weather flow channel of the river are at EL 18 m blocks 48-51, immediately on the right, have their foundation level at EL 24 m, and the blocks 34-40 on the left have theirs around EL 16 to 20 m. The foundation levels of the service spillway blocks 28-33, on left flank, rise gradually, the highest being at EL 47.80 m for block 28. The requirement of conjugate depth for formation of hydraulic jump has necessitated sill level of the stilling basin imperative at EL 12 m. Thus all service spillway blocks with foundation levels higher than EL 18 m have to be provided with the downstream sloping struts, as part of rock slope, from their foundation planes down to bottom of the stilling basin.

Design Approaches

Concrete gravity dams are normally founded on all-rock foundations. Sometimes topographical and hydraulic conditions require them to be constructed partly on rock and partly on sloping concrete strut. The design and analysis of such a sloping strut by conventional methods presents difficulties and several limitations. In this method, the dam portion above the strut would be analysed as a gravity section resting on a complete rigid base and principal stresses at the base would be computed. The strut would be taken as a concrete prop resting on an inclined support with loads computed from the principal stresses acting as surcharge at the top. This approach does not take into account the deformation of foundation rocks or actual site conditions governed by the deformability of rocks and concrete, particularly when the *moduli* of deformation of the rock and the concrete vary to a large extent.

Structural model testing was not adopted. The other method is photo-elastic analysis. In this method the representation of different foundation rocks with different *moduli* of deformation may have its own limitations, although the normal procedure of simulating different *moduli* of deformation in photo-elastic studies is available, using the technique of proportionately reducing the thicknesses of model components.

The realistic behaviour of the dam can be obtained using the finite element method, which takes into account actual distribution of stresses as well as deformability of foundation rocks. For elastic isotropic material, modulus of deformation and Poisson's ratio define material property. Since respective material property could be assigned to a particular element, it is possible to account for non-homogeneity by assigning appropriate material properties to different elements in the structure. The finite element method is, therefore, used in analysis and design of sloping struts.

Constructional Aspects

Ten spillway blocks where sloping struts are provided are Block nos. 28-31, 34-35 and 48-51. The depths of the struts vary from 6 to 30 m. FEM analysis has enabled design of relatively thin and economical sections compared to sections designed by the conventional approach. At its deepest section the minimum thickness of the strut, along with contribution from curvature introduced for transition between the service spillway and stilling basin, is 7 m. Such design would facilitate in totality the proper construction of the struts along with anchors for making them integral with foundation rocks and drainage arrangement at the interface with the rocks. It would also ensure that dynamic forces on the struts due to the hydraulic jump for dissipation of energy of overflowing flood waters in the stilling basin are resisted safely. This objective is fully met with as concrete in the struts would have two grades: A_{40} S_{350} for the outer 1 metre face to safely resist wear and tear and possible cavitational forces due to high velocity flows over the spillway and in the remaining portion of the struts A_{150} S_{160} which is the same as that used for the dam. The concrete for construction of the struts would be pre-cooled concrete with temperature of 12.5°C during placement to avoid its cracking later.

The final configuration of the struts has been adopted with stepped excavation. Controlled blasting and line drilling was resorted to so as to avoid over break in the rock profile supporting the struts. Line drilling was carried out in a single row of 75 mm diameter holes spaced at 2 to 4 times the hole-diameter, drilled along the neat excavation line. These holes were not loaded with explosives. During construction, as found necessary, pre-splitting and delay blasting was at times done to suit the actual site requirements. As the joint pattern was found suitable for providing a steeper slope, stepped excavation with steeper slope of [0.6 (horizontal): 1 (vertical)] with berms at suitable intervals as proposed in the design for the struts was found to be a sound design and favourable to the actual site condition.

The analysis by finite element method has enabled design of relatively thin and economical sections compared to those designed by the conventional approach.

During construction the application of techniques of controlled blasting and line drilling together with pre-splitting and/or delay blasting, wherever necessary, has ensured excavation for the struts to the proposed design slopes consistent with joint pattern in foundation rocks to avoid overbreaks. The use of A_{150} S_{160} and A_{40} S_{350} concrete for the struts and their outer surface respectively would ensure efficient structural and hydraulic performance.

Irrigation Bypass Tunnel (IBPT)

Irrigation bypass tunnel is a normal and a common feature in multipurpose irrigation projects. Its main objective is to ensure continuous flows in the main canals if water flow through the canal-head power house is interrupted. This situation often arises when the power house is under repair, under a forced shut-down, in a situation beyond normal control, or for any other unforeseen reason. In addition to the above, the NWDT award, final orders, provide that whenever there is surplus water in a particular year, the party state can utilise the surplus as available on pro rata basis (shares of the party states). The Narmada Main canal has been designed to carry 1133 cumecs. Accordingly, in the year when surplus water is available over and above the peak requirement for irrigation, which will be passed through the CHPH, balance pro rata share of the surplus water can be passed through the IBPT. The IBPT for SSP was completed in the year 2001-02, so as to provide water to the Main Canal.

POWER HOUSE

River Bed Power House and Canal Head Power House

Both the SSP power houses—CHPH and the River Bed Underground Power House, have been completed in all respects and are regularly generating power. Due to the dam height having not reached the designed height, up to EL.154.00, generation is not at full designed potential.

Garudeshwar Weir

Up to the end of Stage-II, as contemplated in the NWDT Award, MP and Maharashtra would be in the process of developing their irrigation potential. After completion of Stage-III, all the water allocated to both MP and Maharashtra would be fully utilised and as such there would be no surplus water flowing downstream of Narmada Sagar except regulated releases from the hydro-power projects in Madhya Pradesh. To overcome this eventuality, it has been contemplated that the remaining component of the Power House complex, the Garudeshwar Weir with crest elevation at EL 31.75 m, will have to be provided 11 km downstream of the main dam. Since, the turbines installed in the Underground River Bed Power House are Reversible, the weir will facilitate forming a tail pond pool to operate the turbines in pumping mode. Water from the pool will be pumped and restored in Sardar Sarovar reservoir behind the main dam, enabling power production in the RBPH. After Stage-III, and completion of the weir, the power produced by the RBPH would normally be used to fulfil the peak demands—normally three to four hours in the morning and in the evening. During non-peak hours the turbines would operate in pumping mode.

The power production of both the power houses of the SSP until now (30.11.2010) is of the order of 1945 MW at CHPH and 15,720 MW at RBPH. The revenue generated until now comes to Rs. 39.4 billion for RBPH and Rs. 5.0 billion for CHPH. In other words, SSP has already ploughed back Rs. 44.40 billion (amounts rounded off).

4

Seismicity and Safety of the Sardar Sarovar Dam

G.L. Java

This chapter is basically divided into two sections: The first section covers broadly the basic aspects of seismicity and the SSP, so as to understand and comprehend the Aseismic Design for the Sardar Sarovar Dam. Since the SSP has been under criticism from environmentalists for many issue, seismicity and dam safety being one of them, the second section covers the controversies arising out of the writ petition filed by the Narmada Bachao Andolan in the Supreme Court of India.

SECTION I

SEISMO-TECTONIC SETUP

The Sardar Sarovar Dam (SSD) and reservoir are located in a crucial area in the Deccan trap province. The deeper and major earthquakes occur where tectonic plates are pushed together, as in the subduction zone along the Himalayas. The Deccan traps lie well within the Indian plateau, at a considerable distance from the plate boundaries. There are no global stresses being built up here. Thus the probability of major earthquakes (beyond Richter Magnitude 7.0) in this zone is very low.

However, the closest feature of interest for Sardar Sarovar is the Narmada-Tapi-Sone lineament, which some investigators consider seismogenic. Describing the seismo-tectonic environs of the Narmada extending from the Gulf of Khambhat in the west to the Sone Valley in the east, one view is that the river flows along a major fault zone and that the Narmada Valley is an elongated E-W to ENE-WSW trending subsidence zone forming a major rift. But other investigators like Jai Krishna *et al.* (Krishna, 1989) point out that there is no geological evidence that such a fault exists.

The Sardar Sarovar Project (SSP) has been broadly described elsewhere in this book. The present chapter deals with its safety aspects from the Seismicity point of view. A review of the geologic and tectonic setting of Narmada System in general and the SSP site in particular is called for before discussing site-specific seismic parameters.

Regional Geological Setup of Sardar Sarovar Dam

The Pre-Cambrian peninsular shield and the surrounding sedimentary basins (with Tertiary succession) comprise the following:

(i) Cambay basin-West coast off-shore sedimentary basins .... in the West
(ii) Punjab, Ganga/Bengal basins .... in the North
(iii) Mahanadi and Godavari coastal tracts besides east coast basins. .... in the East

The details are shown in Figure 1.

The regional seismo-tectonic map of a part of India focusing on the environs of the Narmada is shown in the Figure 2. The lineament map showing fault disposition around the site as per Geological Survey of India is also available and has been considered. While major faults are normally seen in the main shield regions, relatively young fault zones are seen in the surrounding portion of the shield.

Earthquake data, both historic and recent, collected from various sources indicate that although significant earthquakes occur along the Himalayan belt and the Northern Indo-Gangetic plain, the Indian peninsular shield and its coastal tracts exhibit much lower seismic activity. Seismic activity in the peninsular shield except the surrounding basin and coastal tracts is rated 'below moderate'.

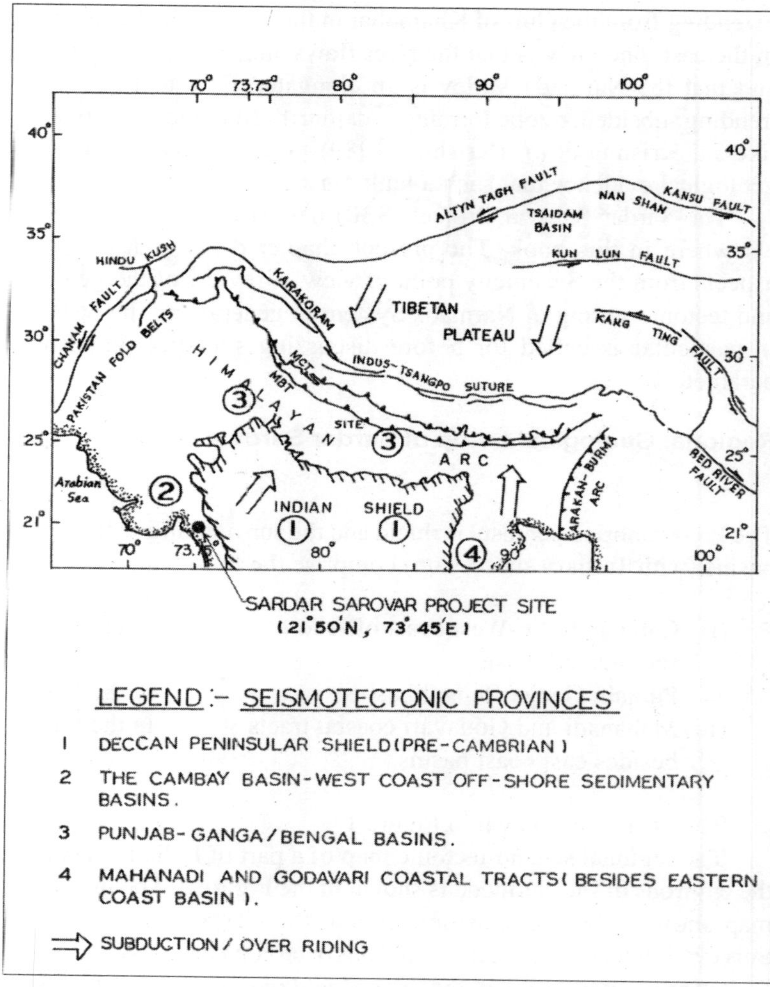

Figure 1: Pre-Cambrian Peninsular Shield and the Surrounding Sedimentary Basins

Earthquakes of Richter magnitude 5 and a few of magnitude 6 and above are associated with these younger fault zones. Besides, the recorded data of earthquakes in the Indian peninsular shield do not indicate any definite associations with known faults and other tectonic lineaments. This geologically ancient stable shield, forming a rigid

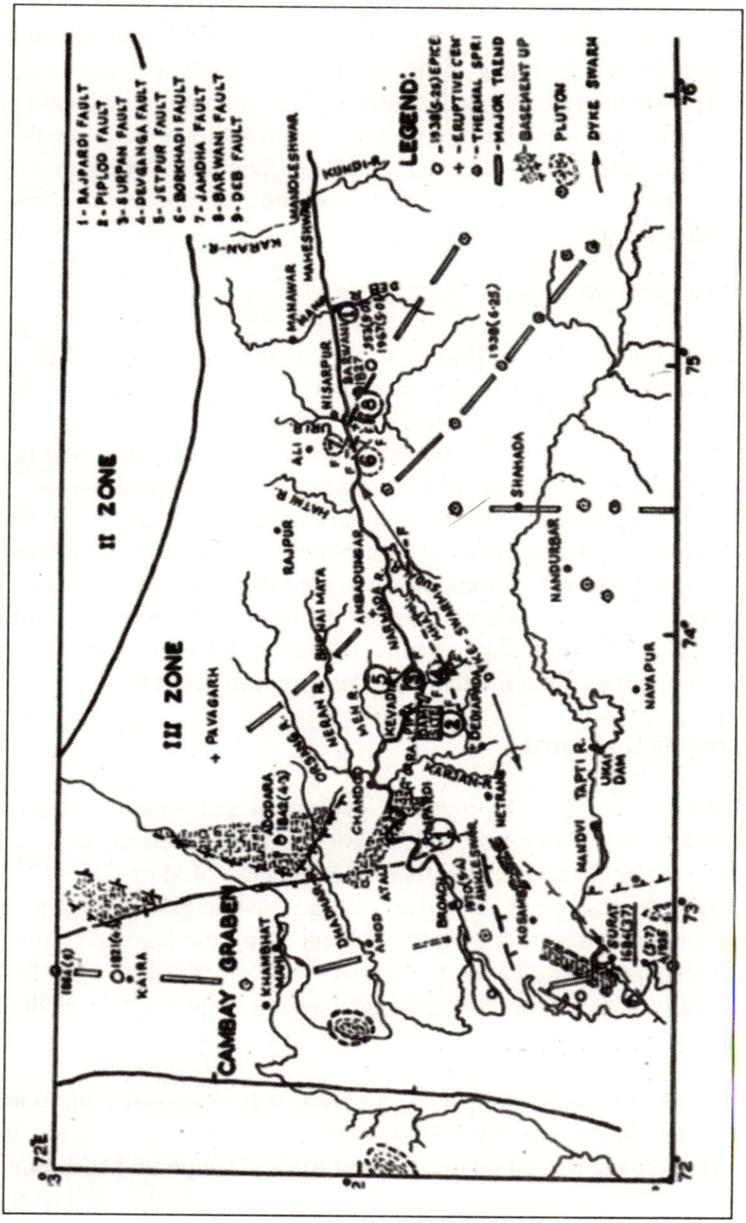

Figure 2: Regional Seismotechtonic Map

continental plate, has undergone major uplifts and subsidence like Gondwana troughs, volcanism along Pre-Cambrian lineaments and other movements resulting in development of faults and fractures. Rejuvenation of pre-existing lineaments in such an interpolate region with a sedimentary basin along its borders can safely be said to pose low seismic potential for large-magnitude earthquakes in comparison to continental plate boundaries, subduction zones, mid-oceanic ridges and island-arc tectonic framework.

Geology at Dam Site

Geological investigations done for the Sardar Sarovar site indicate that different lava flows of Deccan trap and dolerite dykes form the foundation of the dam. These flows, in general, are horizontal. However, they do exhibit local attitude in limited areas with a dip of 5° to 10°. Trap flows overlie a Cretaceous sedimentary sequence, known as Bagh series, which comprises sandstone, indurate shale and limestone. The sequence was intruded later by dolerite dykes and sills. Subsequent faulting has dissected the area and three such fault zones have been identified near the dam site. The main fault along the fair weather flow channel of the river traverses the dam foundation intersecting the dam axis at approximately 80°.

Seismicity Concerns

It is in the context of the regional, geological and seismo-tectonic setup that we have to view the SSD location in terms of seismic activity and safety considerations. The Bharuch earthquake of March 23, 1970 is cited by some experts as evidence of the seismicity of the region. The earthquake is attributed to movement along the Narmada fault line and hence the dam is considered to lie in the heart of the tectonically active zone. Some other questions raised relate to the following:

(i) Gravitational anomaly and crystal subsidence of a narrow belt at the site of the SSD;
(ii) Seepage that could occur along Amba-Dongar and Nisarpur

fault zone which could induce seismic activity; or Reservoir Induced Seismicity (RIS) after filling the reservoir;

(iii) Results of measurements of heat flow spectrum;

(iv) Results of measurements of magnetic polarity of either side of Narmada river and at fault zone;

(v) Surface and underground episodic studies to know the exact nature of spasmodic movements of rocks along the faults after loading the dam with the reservoir;

(vi) Deep seismic studies, and the results thereof, to evolve a three dimensional profile; and

(vii) Experimental modelling to ascertain the future mode of neo-tectonic movement at the dam site.

In fact in many forums and debates including those by Narmada Bachao Andolan (NBA) and other public bodies, in addition to the above points, the question of the safety of the dam against seismicity has been often raised. Hence, for a clear and better appreciation of the various issues involved, it would be appropriate to appreciate the basics of earthquakes and dam design before we consider the various details of relevance. It is earnestly hoped that after going through these clarifications and through the broad details of the writ petition by the NBA in the Supreme Court, it would be clear to all that the SSP and the main dam are safe.

EARTHQUAKES, FAULTS AND GROUND MOVEMENTS

Earthquakes

The planet Earth, born about 4,800 million years ago is sub-spherical in shape, with a diameter of about 12,740 km. Since its birth it is cooling, the surface being the coolest, the core the hottest. The density and temperature differentials set in motion a sort of convection current in the earth's mass. The cooling process also causes cracking in the rigid crust, called the Lithosphere. Its composition is not uniform. Hence cracking takes place in the comparatively weak zones, but with accumulated strains. When such energy is released, the earth's crust shakes and causes earthquakes of tectonic origin. Earthquakes of volcanic origin are rare and India is free of them. A major tectonic

Earthquake is preceded by foreshocks and/or followed by aftershocks, which represent small ruptures of plastic deformation. The capacity for stress accumulation in a rock mass may be a function of its depth of location below the earth's surface. Larger build-ups happen at greater depths, where large magnitude earthquakes can occur.

Collision Boundaries and Faults

Earthquakes are caused on plate collision boundaries, subduction boundaries or on boundaries of closely moving plates. When two plates collide, the one with higher density sinks below the one with lower density. This phenomenon is known as underthrusting or subduction.

Earthquakes also occur within a plate area. Finite sized ruptured surfaces like a discontinuity in a seemingly continuous rock mass are known as 'faults'. A quiescent fault is one which has not been activated for about 10,000 years and it really does not matter. An active fault matters because it could trigger an earthquake nearby. The movement starts at a location, spreads over an area in all or specific directions and stops when the residual strain cannot cause further rupture. The tremor lasts a matter of tens of seconds.

An existing tear or fault provides a line of least resistance for movement and dissipation of accumulated thermal stress/strain. The location, size and recentness of activity along a fault, therefore, are ascertained to predict the location and magnitude of possible further earthquakes. The size of a fault is generally described by its length, which is measured with the help of geomorphic evidence, aerial mapping or remote sensing supported by ground truthing.

Location and Magnitude

The focus of an earthquake is an instrumentally determined area below the earth's surface where slipping or fracture begins. Magnitude is a rating of a given earthquake independent of the place of observation. It is calculated from measurements on seismographs and expressed in ordinary numbers and decimal fractions based on a logarithmic scale.

It is also common to indicate the effects of earthquake on an arbitrary intensity scale like Mercalli. While the magnitude of an earthquake does not indicate its damaging effect, the intensity scale, expressed in Roman numerals, does.

An idealised relationship between magnitude and the length of the fault which undergoes rupture can be indicated. This provides an approximate rupture length at source and energy release, corresponding to the magnitude (Narmada Control Authority, 1993).

Ground Movement, Peak Ground Acceleration and Response Spectrum

The primary effect of an earthquake on any structure is to shake it at ground level for a short duration—normally less than a minute. The tremor can be idealised as the longitudinal, transverse or vertical movement of ground particles. This movement is around a stable position and is described by velocity or acceleration around it. Gravity acceleration (*g*) being a universally accepted standard, the ground acceleration during each cycle of vibration can be described as its fraction, such as 0.05*g*, 0.15*g* etc. meaning that the acceleration force of the cycle was 5 per cent or 15 per cent of the gravity acceleration respectively.

The vertical acceleration is normally less than the horizontal acceleration, except in near source record, and is of the order of about 50 per cent to 75 per cent. For design of a structure, the vertical acceleration and the higher of the two values in horizontal plane are significant.

The effect of earthquake tremors is captured in an accelerogram, which is a graph of the tremors showing the acceleration caused to the instrument recorder through every cycle of vibration. The peak value of the event is known as peak ground acceleration (PGA). The energy content of an event which affects a structure is best given by the integrated effect of each cycle of vibration over the period of shaking and not by the PGA alone.

Accelerogram records from various sites around the world show that the largest peaks of acceleration on the record may occur in only a few spikes. It is now common practice to evolve and utilise for the design of structures a representative value known as Effective Peak Ground Acceleration (EPGA). This value is the one that is statistically relevant and appropriate for risk analysis and design purposes.

A response spectrum is a plot of maximum values of acceleration, velocity and displacement responses of an infinite series of single-degree-of-freedom system subject to time-dependent dynamic excitation. The maximum response values are expressed as a function of undamped natural period for a given damping.

Maximum Credible Earthquake (MCE)

International dam design practices recommended detailed dynamic analysis for all dams with high and extreme risk rating. For this purpose, it is necessary to evolve seismic design parameters based on a proper seismo-tectonic investigation. The procedure involves identifying appropriate levels of earthquakes which can be justifiably assigned to these sources.

A notional MCE can be associated with each seismogenic source such as a fault system. This is the largest reasonably conceivable earthquake that can happen near the source within the geographically defined tectonic province and framework. It is also defined as an upper bound of expected magnitude. The Maximum Design Earthquake (MDE) will normally be characterised by a level of motion equal to that expected at the dam site, from the occurrence of deterministically or probabilistically evaluated MCE. The Operating Basis Earthquake (OBE) represents the level of ground motion at the dam site at which minor damage to the structure is acceptable. The OBE is determined by probabilistic methods. A critical MCE (CMCE), however, is worked out to envelope MCE from different identified seismogenic sources. The value of Design Basis Earthquake (DBE) is usually taken as half of MCE. In structural design, it is normally ensured that the MCE should not cause a concrete dam to crack, slide or become functionally impaired.

SEISMIC PARAMETERS FOR SSD

Seismic Sources

Earthquake data for peninsular India show that the highest magnitude of an MCE can be 6.5. The fault and tectonic lineaments of relevance for SSD are as follows:

(i)	Rajpardi and associated faults forming the eastern boundary of Cambay basin West Coast seismo-tectonic province.	Shortest distance to the lineament from dam site is 55 km towards west.
(ii)	Barkhadi-Barwani fault	Shortest distance from site is 110 km east.
(iii)	Piplod fault	Shortest distance 12 km south of dam site
(iv)	Tilakwada-Bardoli fault	Shortest distance 17 km north of dam site.

While it is plausible to associate a Richter magnitude 6.5 earthquake only with cases (i) and (ii) above, which are the only seismogenic lineaments that are capable/active, a 'worst case scenario' of a Richter magnitude 6.5 earthquake has been postulated at an epicentral distance of 12 km with a depth of focus of 18 km for case (iii), for conservative design estimates of ground motion characteristics.

River Bed Fault

The prime question that arose, apart from the seismicity of the dam area in general, was related to the presence of a river bed fault. The Geological Survey of India (GSI) ruled out ground movements at this fault. Based on an in-depth investigation, the following findings emerged:

(i) The entire Deccan trap region from Netrang to Barwani does not show pronounced activity in historical times. The faults are considered innocuous.

(ii) The local river/channel fault at the dam site was established as an *en echelon* fault, shear zone and without original extension. The fault is bounded by the Mokhadi fault upstream and Akalbar fault downstream with a total length of only 2 km. It was not considered a part of the Narmada-Sone lineament which lies about 17 km north of the dam site. No visible evidence of recent movement along this fault was observed. A fault or lineament of a length 15 km or more is normally required to actuate a Richter 6.5 magnitude earthquake. Since the river channel fault was of 2 km length only, it was not considered as a candidate to cause a credible earthquake of significance for design.

(iii) Central Water and power Research Station, Pune, and School of Earthquake Engineering, University of Roorkee both undertook a micro-seismic survey in the area in 1970-75 and 1980 respectively and confirmed that the river bed fault crossing the dam site registered no activity.

(iv) An International Dam Design Review Panel, comprising eminent and experienced experts on geology, seismology and dam design, considered all the relevant facts/reports and field data and expressed no reservation about the established inactive nature of the river bed fault and adequacy of tectonic investigations.

Different views on the status of the fault have been covered already in the other chapter appearing in this book that has been referred to above.

The consensus reached after various discussions was that there is no evidence at present to presume that the riverbed fault is active. It may however be prudent to assume that the Narmada-Sone lineament is seismically active and to design the project structures accordingly. While doing this it may also be noted that a seismic event would have two aspects—vibrations due to shock and physical relative movements along the fault plane. Design to withstand the vibrations was done for an agreed factor of 0.125*g*. For the physical movements it was noted that the length of the fault will have to be sufficiently long to admit a sizable movement. Therefore, the possibility of significant movement along the fault plane met within the seat of the dam is remote as it peters out within 500 m upstream of the dam and its length is about 2 km only. This is very clear and evident from the discussions in sub-para (ii) above.

Accordingly, after going through all these views expressed by various experts, the DDRP was of the opinion that the River Bed Fault should not be a matter of concern. However, adequate precautions and factor of safety should be built in while finalising its design parameters.

DESIGN OF DAMS FOR EARTHQUAKE FORCES

Design Process

The dam design process is evolutionary. Refinement is undertaken step by step, compatible with the stage in design. It involves rigorous, complicated and time-consuming procedures, which require characterisation for dynamic properties of materials/foundations and adopting them in numerical analysis.

While the first stage in design proceeds with a simple methodology like the pseudo-static method, in the design of concrete dam checked by simplified dynamic analysis procedures evolved by some authors like Chopra and Saini, a detailed analysis for stress distribution and stress concentration in the dam section is carried out using refined techniques in the second stage. Minor modifications in the dam section, if warranted, are carried out during the construction phase to improve the structural behaviour.

Pseudo-Static Method of Analysis

The preliminary design of section is evolved using a 'seismic coefficient' that reflects appropriately the earthquake forces anticipated in the particular region. To facilitate the procedure, the country is divided, as per BIS Code IS: 1893-1984 into five zones with assigned seismic coefficients. These coefficients account for (i) Importance of structures, and (ii) Soil/foundation conditions.

The approach known as pseudo-static analysis aims at converting dynamic earthquake forces into equivalent static forces and applying them in the direction of destabilising forces to calculate the factor of safety. The approach is widely used because of its simplicity, consistency and ease of comprehension in terms of adequate factor of safety.

DYNAMIC ANALYSIS/ADVANCED PROCEDURES

Seismic Parameters for Design

The preliminary design is test checked during detailed design phase for site-specific seismic parameters adopted after detailed studies.

Such studies proceed in two phases. In the first phase, site-specific seismic design parameters are evolved. For this purpose, geo-tectonic and seismic data are assembled and analysed to establish seismogenic sources and credible earthquakes that can be associated with each one of them. The MCE is the severest one which could cause maximum ground acceleration at the dam site, given a set of geo-tectonic features and attenuation characteristics of the formations through which the seismic waves travel. The PGA established by this methodology is different from the 'generalised' seismic coefficient. A lot of confusion has resulted from the mixing of 'seismic coefficient' with 'peak ground acceleration'. In order to cover the seismic risk, what is of interest to designers is a value of PGA that is relevant in a statistical sense. In other words, stray peak values are excluded and an EPGA is evolved that is representative in character. It corresponds to the MCE at zero period acceleration in the response spectrum where that spectrum considers vibration characteristics inherent to the structure as reflected by its natural period of vibration.

An artificial design accelerogram compatible with the response spectrum is normally derived for the purpose of detailed analysis, using a step-by-step method that incorporates in-depth understanding of the dynamic behavioural stresses, strains and deflections.

Institutional Base for Aseismic Design of Dams in India

As the subject of earthquakes and dam design cover various disciplines of science and technology, the institutional support in our country has been broad based for different aspects.

Investigations

At the Government of India level, the subject of geology is dealt with by the Geological Survey of India (GSI); seismologic observations are dealt with by India Meteorological Department (IMD); the Central Water and Power Research Station (CWPRS) deals with seismological investigations related to engineering structures; besides, the Department of Earthquake Engineering, University of Roorkee is the premier institution assisting in studies related to assessment of aseismic design parameters and aseismic design of structures. In addition, through specific projects, National Geophysical Research Institute (NGRI), Wadia Institute of Himalayan Geology (WIHG), National Remote Sensing Agency (NRSA), Survey of India (SOI) and the Department of Science and Technology (DST) contribute on the subject.

National Committee on Seismic Design Parameters (NCSDP)

In the year 1969, the Ministry of Water Resources (MOWR) set up a Standing Committee for determination of Design Seismic Coefficient with experts drawn from these institutions to advise dam owners/ projects about the seismic design parameters to be adopted for dams. The Committee, subsequently renamed as National Committee on Seismic Design Parameters (NCSDP), has so far advised agencies of more than 125 dams regarding seismic design parameters. Utilising such advice, seismic coefficient was suggested as 0.125*g* for preliminary design, and the basic dam sections were evolved by the State Design Engineers. Later, specific studies were commissioned to evolve the Seismic Design Parameters through Department of Earthquake Engineering, University of Roorkee. Based on a report and suggestions obtained from these studies, work on the detailed designs proceeded.

Structural Design

The Central Water Commission (CWC) is the apex organisation in Water Resources Sector with a strong in-built Design and Research (D&R) Organisation. It provides guidance to all the state D&R Organisations besides rendering consultancy for specific jobs. Some Indian Institutes of Technology and Engineering Colleges spread throughout the country are also engaged in consultancy work in specific cases on aseismic design of dams.

The unique situation in the State of Gujarat has been that a well-knit engineering organisation exists to handle dam projects. Experts who handle design, research, and quality control as well as construction aspects work together with an able band of engineering geologists. A separate team has been carved out to handle SSD considering the colossal size of the works involved.

Most of the National Agencies were involved in various stages for the consideration of aseismic design of SSD. GSI was involved in seismo-tectonic investigations since very early stages. Studies to evolve seismic parameters were undertaken in the Department of Earthquake Engineering, University of Roorkee by a Group of Experts in different fields like geology, earthquake engineering and civil engineering. Besides the state's Design Cell which took up pseudo-static design, experts from Roorkee University carried out dynamic analysis. The designs were scrutinised by an international Dam Design Review Panel. A Design Cell was created in CWC for vetting the dam design. Thus it would be seen that the project had full infrastructural capability to evolve a safe and sound design of dam and other components.

SARDAR SAROVAR DAM SECTION

Process of Evolution

The process of evolution of the SSD Section to withstand destabilising forces proceeded in stages, from preliminary design to final design. For the preliminary design stage, an a-seismic design was evolved using a seismic coefficient of 0.125*g* as approved by NCSDP. The design seismic coefficient adopted in this procedure is different from PGA. The seismic risk covered by this value is such that many peaks of an actual accelerogram can exceed it several times.

The detailed design of the dam section for SSD is based on the EPGA of 0.25*g* though a lower value of 0.21*g* was originally obtained through empirical procedure. The value of 0.25*g* for EPGA was considered in depth by the Dam Safety Review Panel. The project had consulted R.W. Clough of University of California, Berkeley, who in turn consulted Bruce A. Bolt of the same university and endorsed the adoption of EPGA of 0.25*g*. The response spectrum using this as zero period acceleration and an artificial design accelerogram compatible with the response spectrum were duly considered. This accounts for the recorded accelerogram of the Koyna earthquake by appropriate scaling up by a factor of 1.2 over the Koyna accelerogram to take care of the SSD site geology.

Thus all conditions to evolve an appropriate accelerogram for a safe and sound design of dam section were fulfilled.

Design of Section

The project authorities resorted to the standard methodology in arriving at a safe design section for SSD. A typical cross-section of a dam is shown below:

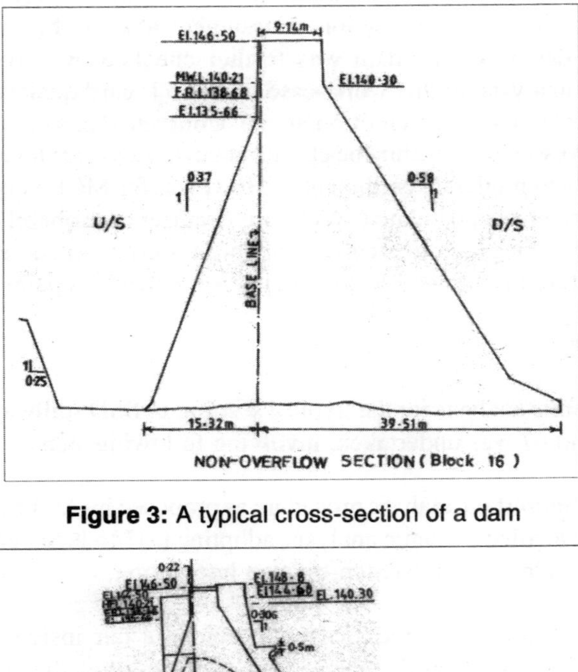

Figure 3: A typical cross-section of a dam

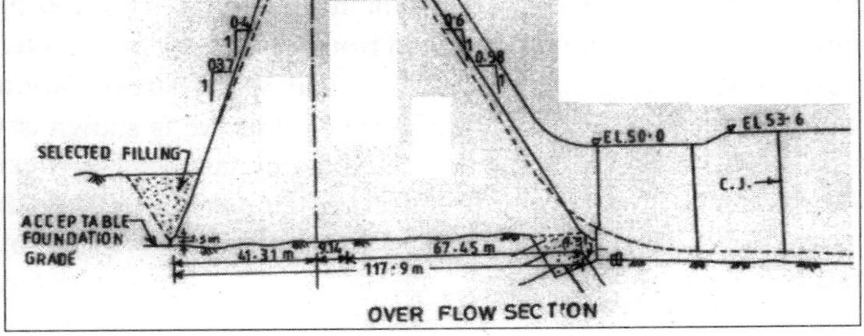

Figure 4: Over Flow Section

Typical Sections of a Dam

The preliminary design of the dam was done using IS provisions with a fairly severe combination of other forces. The IS code stipulates that the dam should have no tension under normal loading condition (IS:6512-1984) and permit nominal tensile stresses for extreme loading conditions. The section as designed satisfied this criterion.

The design of the dam was further checked by carrying out dynamic analysis for the 'worst-case scenario 1' earthquake allowing for the expected ground motion at site. Concrete dams are designed to remain essentially within the elastic range of behaviour for relatively frequent and moderate earthquakes. However, for MCE very intense ground tremor is postulated. With rare occurrence probability, some limited cracking is tolerated which would not impair the ability of the structure to impound water and is economically repairable.

Analysis

The dynamic analysis for the 'typical section' of SSD spillway Blocks 34, 43 and 47 was undertaken, involving following steps:

(i) Simplified analysis procedure as proposed by Anil K. Chopra.
(ii) Detailed dynamic analysis, adopting (a) Mode Superposition Method, and (b) Step-by-step Integration.

The detailed dynamic analysis provided a fair insight into the extent of stresses, both tensile and compressive, that could be anticipated for different loading situations. A proper insight into the type of results obtainable from dynamic analysis would require scanning scores of printouts obtained from a computer-run, each one for different loading conditions. A typical set of stress variation contours in respect of Block 43 of SSD earthquake is shown in the Figure 5. This is based on the detailed dynamic analysis carried out in Department of Earthquake Engineering, University of Roorkee.

Results

The results obtained from dynamic analysis as well as by conventional pseudo-static methodology proved that the stresses (including

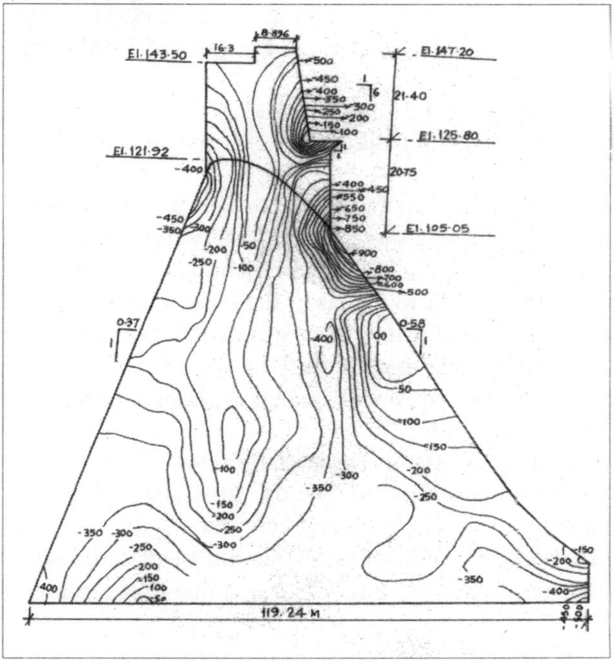

Figure 5: Principal Stress Due to Static Load+EQ Load (Block 43), Sardar Sarovar Dam

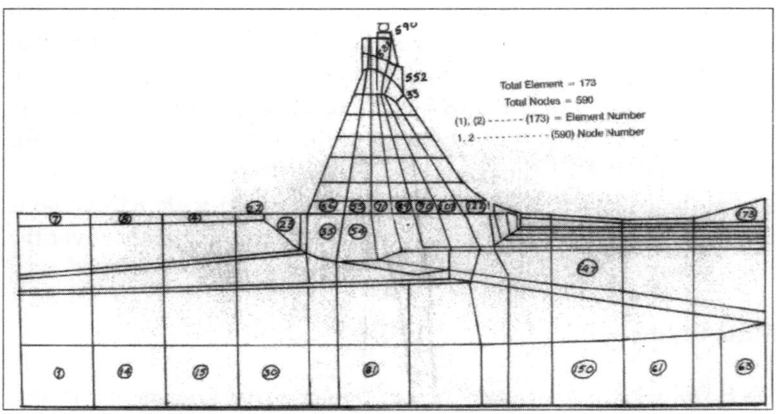

Figure 6: Finite Element Mesh of Block 43with Pier

tensions) during the MCE are within acceptable range and at acceptable locations. These regions are provided with appropriate structural safeguards.

Reservoir Induced Seismicity

The increase in native seismic activity associated with the filling of some large reservoirs, documented in recent years, led to an apprehension about likely RIS at SSD. Though debated extensively, the largest recorded RIS to date is of the order of 6.3 in India and 6.5 elsewhere in the world. Most instances of RIS reported all over the world have much lower magnitude.

The primary mechanism advocated explaining RIS is as follows:

(i) Effect of shear stress created by reservoir load;
(ii) Added stress effect of already 'fully stressed mass';
(iii) Increased pore-pressure effect; and
(iv) Decrease in coefficient of friction of material such a fault gouge, if present, when saturated due to reservoir.

It is significant to note that direct relationship between the mechanism and RIS has not been established clearly. Nevertheless, the induced stress changes associated with reservoir impoundment are small when compared to tectonic stresses and stress drops associated with large earthquakes. In fact, it has been viewed that RIS results in gradual and progressive release of ambient or residual stresses along pre-existing zones of weakness. It has also been seen that the maximum RIS does not exceed the probable tectonic earthquake. In a situation where the tectonic activity accounted for in SSD has a Richter magnitude 6.5 along the closest fault system and as the generally accepted upper bound world-wide RIS value is also 6.5, the design of the dam is considered to adequately cover the risks, it any, due to possible RIS.

MEQ Studies

Such studies are under taken to decipher active lineaments, if any, the scale of present activity and to enable designers account for it appropriately.

The Department of Earthquake Engineering, University of Roorkee, undertook detailed studies around the SSD site between February-August 1980. Six months of readings were taken at five stations and the data obtained on micro earthquake recordings processed. The following conclusions emerged from the studies:

(i) The river bed fault, which was not loaded during the period of micro-earthquake recording, did not show any seismic activity.
(ii) While several events were registered and located in and around Rajpardi fault in the 1976 micro-earthquake survey, the 1980 record showed complete absence of activity from this fault.
(iii) The highest magnitude of event recorded during the survey was less than 3 on the Richter scale. Activity observed near the Sone-Narmada lineament running ENE/WSW and the Piplod fault could have focal depth in excess of 15 km.
(iv) The micro-earthquake activity provided enough data for statistical extrapolation for periods up to even 100 years. The magnitude of maximum earthquake extrapolated works out to less than that for the design earthquake, which is considered conservative.

The investigations to determine the seismic design parameters for SSD involved established authorities of the country. National and international standards were followed. Theoretical techniques of dynamic analysis and state-of-the-art technology were used to assess the behaviour of the designed dam section. For scrutiny and vetting of designs, advice was taken from nationally and internationally reputed consultants to ensure that the dam would serve the best interest of beneficiaries for centuries even under the worst earthquake.

The SSD design and sections have been evolved and analysed meticulously by competent experts and debated by panels of renowned and experienced professionals. All safety requirements are ensured as per the requirements. It will be seen from above that fears about adequacy of the dam design for a severe earthquake are thus ill-founded and should not cause concern to anybody.

SECTION II

Apprehensions Regarding Safety and Vulnerability of the Main Dam Due to Seismicity

Despite all safety requirements being fulfilled as discussed above, in keeping with BIS and international standards, including good engineering practices and other seismicity provisions, fears about the adequacy of the dam design to withstand a severe earthquake were still being expressed. Notwithstanding the assurances, the NBA filed a writ petition in the Supreme Court of India.

This section covers broadly the main aspects put forth to the honourable Supreme Court in the Writ Petition (Civil) No. 319 of 1994 filed by the Petitioners Narmada Bachao Andolan (NBA) *vs.* Union of India and others, and the reply to the supplementary affidavit of the petitioners regarding the matter of the Jabalpur Earthquake and its implications for SSP. In fact, earlier also writ petitions had been filed at different times by the NBA on seismicity and other issues. In this section, I have focused on the seismicity aspects of the writ petition and the proceedings.

The primary concern of the NBA in this case was that as the dam was not safe and adequately designed for the Safety against Seismicity, in the event of its failure, it could be a manmade source of death and destruction for many humans and large amounts of property in the entire river valley downstream of the dam. Thus the Judicial provision for the well-being of society at large and the general safety of the public in particular, to have a safe and protected sound social system was understood.

All the writs filed and debated by NBA, whether on this particular topic or on other topics in the Honourable Supreme Court lasted for a period of almost six long years from 1994 through 2000. These are also broadly discussed and covered below.

In the Supreme Court of India (Civil Original Jurisdiction) Writ Petition (Civil) No. 319 of 1994 was filed by the Petitioners Narmada Bachao Andolan (NBA) *vs.* Union of India and others…Respondents was filed. While this petition was still being heard in the Supreme Court, I filed an affidavit in reply to a supplementary affidavit regarding the matter of the Jabalpur earthquake and its implications for the Sardar Sarovar Project. Some relevant details of the said affidavit are reproduced below:

"In the Supplementary Affidavit of 17th July, 1997*, the petitioners have addressed in the main, the following issues, which apparently have been considered of importance by them in the present context:

I. Narmada Valley is the second most important rift zone in the country; and has become more vulnerable to earthquake and is reactivating.

II. Dams being built in Narmada valley are prone to Seismic Risks and if the warnings are ignored, devastation and destruction in the valley are possible.

III. There is a possibility of the seismic activity being triggered off by the dams themselves—the scientific term being "Reservoir Induced Seismicity" OR RIS. They have quoted the "KOYNA DAM" as an example for the largest RIS in the world and have also referred to Sukta Dam (Narmada valley) and Bargi Dam being the cause for RIS (Jabalpur Earthquake).

IV. Even if the dams are constructed is made safe; the seismicity induced by the reservoir can play havoc with the surrounding areas. Cumulative impact of dams considered together has not been evaluated.

V. Repair and restoration cost in the affected areas will be running into several thousand crores of rupees.

VI. Comprehensive Seismic review of the valley including that of the Sardar Sarovar Dam need be done.

VII. Seismic risk evaluation and proper hazard assessment be made.

VIII. Rim stability studies and water loss problem need to be addressed.

IX. Representation to Honourable Prime Minister." (Supplementary Affidavit 1997)

---

*Before filing this affidavit, we had to discuss its technical aspects with our senior-most counsel, Fali S. Nariman, and apprise him of the matter. I was required to brief him on all the technical aspects in the matter of Seismicity of the Sardar Sarovar Dam and the processes for the design and safety adopted to safeguard the structure. He enquired why the Dam Safety Chairman/members could not file this affidavit on behalf of the Government, instead of this author, who explained that as a Chief Engineer Dam and Power House Designs and the highest authority in the Nigam and as a Technical Design Head, he was the only person authorised, as far as this Affidavit was concerned to officially sign and file it on behalf of the Government.

We then discussed practically all the aspects as to how we had designed the structure, what authorities had been consulted or had contributed in finalising its design, which Institutes/Organisations were involved both in India and abroad. I started with the fundamental specifications and provisions of the Bureau of Indian Standards, their requirements including the Seismicity Codes of Practices followed in India and abroad, and briefed him on our discussions with experts in the field of Seismicity in India and abroad. I proceeded to apprise him of the consultations held from time to time with, and approvals taken from, practically all the noted authorities in the Geological Survey of India and other prestigious institutions connected with the seismicity, structural design, geology and other related aspects. I informed him how, finally, the entire design was also cleared and approved by the DDRP and the DSP. This exercise and briefing took six rigorous sessions until he was fully satisfied in all respects.

The writ petition had outlined nine major issues as enumerated above. All these issues were duly replied to and discussed in the said affidavit in the Supreme Court. For the purpose of the present chapter, only the most important first three points are broadly discussed here. Before embarking on the discussion, however, it is necessary to refer to the Indian Standard IS:1893-1984 "Criteria for Earthquake-Resistant Design of Structures" (Fourth Revision), by Bureau of Indian Standards (previously the Indian Standards Institution), brought out in June 1986.

This Standard was adopted by the (then) Indian Standards Institution on 16 November, 1984, after the draft finalised by the Earthquake Engineering Sectional Committee had been approved by the Civil Engineering Division Council. Some relevant extracts from the Standard are reproduced below:

> QUOTE
> "This Indian Standard (Fourth Revision) was adopted by the Indian Standards Institution on 16 November, 1984, after the draft finalized by the Earthquake Engineering Sectional Committee had been approved by the Civil Engineering Division Council.
>
> ....
>
> "Himalayan-Nagalushai region, Indo-Gangetic plain,

Western India, Kutch and Kathiawar regions are geologically unstable parts of the country and some devastating earthquakes of the world have occurred there. A major part of the peninsular India has also been visited by strong earthquakes, but these were relatively few in number and had considerably lesser intensity. The earthquake resistant design of structures taking into account seismic data from studies of these Indian earthquakes has become very essential, particularly in view of the heavy construction programme at present all over the country. It is to serve this purpose that IS: 1893-1962 'Recommendations for earthquake resistant design of structures' was published and subsequently revised in 1966.

....

"It is important to note that the seismic coefficient, used in the design of any structure, is dependent on many variable factors and it is an extremely difficult task to determine the exact seismic coefficient in each given case. It is, therefore, necessary to indicate broadly the seismic coefficients that could generally be adopted in different parts of zones or the country though, of course, a rigorous analysis considering all the factors involved has got to be made in the case of all important projects in order to arrive at suitable seismic coefficients for design. The Sectional Committee responsible for the formulation of this standard has attempted ... to classify the area of the country into a number of zones in which one may reasonably expect earthquake shock of more or less same intensity in future. The modified Mercalli Intensity ... broadly associated with the various zones is V or less, VI, VII, VIII and IX and above for zones I, II, III, IV and V respectively. The maximum seismic ground acceleration in each zone cannot be presently predicted with accuracy either on a deterministic or on a probabilistic basis. The design value chosen for a particular structure is obtained by multiplying the basic horizontal seismic coefficient for that zone ... by an appropriate Importance Factor... Higher value of importance factor is usually adopted for that structure, consequences of failure of which are serious.

> However, even with an importance factor of unity, the probability is that a structure which is properly designed and detailed according to good construction practice, will not suffer serious damage". (Bureau of Indian Standards 1986)
>
> (UNQUOTE)

In the case of major River Valley Projects in the country, in addition to the stipulations of the Bureau of Indian Standards, the design seismic coefficients are recommended by the Standing Committee. This Standing Committee for advising design seismic co-efficient for all the river valley projects has been specially constituted by the Ministry of Irrigation and Power, Government of India.

> (QUOTE)
>
> "After lot of deliberations, the Members of the Committee agreed unanimously, that the horizontal seismic coefficient of 0.125*g* may be adopted for Sardar Sarovar dam. This was specifically done keeping in view the importance of the structure, danger potential and the likely induced seismicity. The coefficient recommended above is for the main masonry dam. The coefficient for the rockfill dams which are smaller in height may be adopted suitably by the project authorities themselves." (Supplementary Affidavit 1997)
>
> (UNQUOTE)

Further, the 15th meeting of the Standing Committee for advising seismic design coefficient held in October 1981, on review of the seismic coefficient for Sardar Sarovar Project, stated as under:

> (QUOTE)
>
> ".... After further discussions in which ... Director, GSI, Western Region and other members participated, it was decided that there is no need to change the earlier recommendations of seismic coefficient of 0.125*g*." (Supplementary Affidavit 1997)
>
> (UNQUOTE)

At the initial design stage of the Sardar Sarovar Project, the Indian Standard IS: 1893-1975 (Third Revision) specified a design seismic coefficient equal to 0.08*g* for zone III, in which the Sardar Sarovar Dam is located. The Standing Committee, Government of India recommended the design seismic coefficient of 0.125*g*, on the conservative side, keeping in view the importance of the structure, its danger potential and the likely induced seismicity. Adopting a higher seismic coefficient results in a safe design of dam section to withstand earthquake forces.

The first three issues of the petitioners are dealt with point-wise herein below:

Issue No. I

Regarding: "Narmada Valley is the second most important rift zone in the country and has become more vulnerable to earthquakes and is reactivating."

The project authorities have been aware of the seismic sensitivity in the Narmada Valley ever since the early stages when SSP was originally conceived. Comparison between the seismic sensitivity of the Narmada Valley and that of the Himalaya is not quite relevant for SSP. The Himalayas are young mountains still in the stage of developing, whereas Narmada Valley is a part of the original stable Indian plate. Seismic events recorded in the Narmada Valley during historical time suggest moderate activity. There is an increasing awareness of seismicity in the present era and more and more data of recorded seismic events both nationally and internationally are becoming available. This, by no means, need be interpreted as an increasing seismic activity. It is essentially a case increasing awareness and recording of data on an increasing scale for designing structures resistant to expected forces of earthquakes in any region. For the Narmada valley particularly in the vicinity of Sardar Sarovar Project, the maximum anticipated magnitude would be 6.5 on the Richter scale, termed as the Maximum Credible Earthquake (MCE) in engineering seismology, as would be evident from the subsequent paragraphs. This has been duly taken care of in the design of Sardar Sarovar Dam in a very conservative manner.

The project authorities were even aware of the Seismic Sensitivity of the region at the initial formulation of the project. This is evident from the extract reproduced below from Chapter XII (Geological and seismological Aspects of Sardar Sarovar Dam site) of Report of the Narmada Water Disputes Tribunal, Volume-II.

> (QUOTE)
> "Apart from the investigations made by the Geological Survey of India, the contending States obtained geological reports on the dam site from their own experts. Madhya Pradesh has exhibited a report dated January 1976, (MP-789) ... based on Earth Resources Technology Satellite imagery and another one ... (Ex MP-839) .... Maharashtra has filled a report (MR-139) ... in 1975. Gujarat has exhibited a report dated December 1976 ... (Ex G-1073). ... Director, G.S.I. has given his comments on the report ... in the later part of 1976 (G-1061). Besides these a number of reports and documents by other scientists have been produced in evidence." (NWDT 1978: pp 53)
> "Considering that the Geological Survey of India is the official agency which is entrusted with all geological investigations of major river valley projects and all such projects are planned on the basis of its findings, we accept the views of Shri Srivastava in the present case regarding the feasibility of the Navagam Site 3 for treatment of the various adverse features met in the dam foundation will be taken at the proper stage by the consulting committee of engineers and geologists referred to by Shri Srivastava." (NWDT 1978: 56)
>
> (UNQUOTE)

Further in this regard, an extract given below from the detailed studies carried out in January 1982 by the Department of Earthquake Engineering, University of Roorkee for the Sardar Sarovar Dam on "Design Earthquake Parameters for Navagam Dam Site in Gujarat" Earthquake Engineering Studies EQ 81-1 (Revised), states that—

> (QUOTE)
> "Seismotectonic Setup: Gubin (1969), Tipnis and Srivastava (1969), Auden (1969, 1975), Balasundaram *et al.* (1970),

> Das and Ray (1976) have described the seismotectonic frame work of the environs of Narmada River extending from Gulf of Cambay in the west to the Son Valley in the east. Most of these views are based on the premise that Narmada Valley represents an elongated E-W to ENE-WSW trending zone of subsidence forming a major rift. ....
> Sardar Sarovar Project site lies in the Pre-Cambrian Peninsular Shield Seismotectonic Province. The Cambay Basin—West Coast Seismotectonic Province with its eastern boundary at about 55 km west of dam site has shown moderate seismic activity in the past. However, only a magnitude 5.4 (1970 Broach earthquake) was observed during recent times. This belt can be taken as a seismotectonic lineament capable of generating magnitude 6.5 earthquakes. Within the Peninsular Shield Seismotectonic Province in the Narmada environs the major tectonic features in the vicinity of the dam site are Borkhadi-Barwani faults located about 110 km east of dam site, ENE-WSW trending Piplod fault about 12 km south of the dam site and Tilakwada-Baroli (ENE-WSW trend) faults about 17 km north of the dam site. The Borkhadi-Barwani faults along which past earthquake occurrence are observed can also be considered as capable faults. No direct evidence of recent movements or association of earthquake occurrence has been reported along Piplod fault and Tilakwada-Baroli faults." (Department of Earthquake Engineering 1982: 7-10)
>
> (UNQUOTE)

These studies also show that this belt can be taken as a seismotectonic lineament capable of generating an earthquake of magnitude 6.5 on the Richter scale. In this context, a list of seismic events that have occurred in the Narmada River region is given below.

Immediately after the Bharuch earthquake of March 1970, two seismological observatories, one on the northern bank of the Narmada near Broach and the other on its southern bank near Ankleshwar were installed in 1970. In 1972, a seismological observatory was installed at the Project colony near the Navagam dam site. Of the earthquakes recorded by this network, those of magnitude 3.5 and above which

List of Epicentres of Important Earthquakes of the Narmada River Region

Year	*Month*	*Date*	*Epicenter*		*Magnitude*	*Area*
			Lat. N	*Long. E*		
1	2	3	4	5	6	7
1684		–	21.2	72.8	3.7	Surat
1821	August	13	22.8	72.7	4.3	Kaira
1827	May	27	– – –			Near Barwani
1842	October	9	– –		4.3	Baroda
1863	November	18	22.0	75.0	5.0	Nimar and Barwani but not noticed south of Satpura range.
1864	April	29	23.0	72.7	5.0	Ahmadabad
1935	July	20	21.0	72.8	5.7	Dumas
1938	March	14	21.6	75.0	6.25	Khandwa-Satpura range
1967	January	6	22.0	75.0	5.0	Dhar and Jhabua
1970	March	23	21.7	73.0	5.4	Broach
1970	August	30	21.6	72.7	3.5	Aliabet

Source: Department of Earthquake Engineering 1982.

occurred in the Narmada basin within 200 km from Navagam, while the frequency of earthquakes of magnitude 3 and above considered to be of engineering significance, which occurred in the Narmada basin within 200 km from Navagam dam.

Issue No. II

Regarding: "Dams being built in Narmada Valley are prone to seismic risks, and if the warnings are ignored, devastation and destruction in the valley are possible."

Before going into this issue, it would be necessary to discuss, how major dams are normally designed. In this regard an extract reproduced below from a reference book Design of Gravity Dams: A Water Resources Technical Publication, United States Department of the Interior, Bureau of reclamation, Denver, Colorado, 1976, indicates the procedure for the Design of Large Dams for Earthquake response.

(QUOTE)

"Design Data and Criteria D. LOADS

Earthquake: Concrete dams are elastic structures which may be excited to resonance when subjected to seismic disturbances. Two steps are necessary to obtain loading on a concrete dam due to such a disturbance. First, an estimate of magnitude and location must be made of the earthquake to which the dam will be subjected and the resulting rock motions at the site determined. The second step is the analysis of the response of the dam to the earthquake by either the response spectrum or time-history method.

Most earthquakes are caused by crustal movements of the earth along faults. Geologic examinations of the area should be made to locate any faults, determine how recently they have been active, and estimate the probable length of fault. Seismological records should also be studied to determine the magnitude and location of any earthquakes recorded in the area. Based on these geological and historical data, hypothetical earthquakes usually of magnitudes greater than the historical events are estimated for any active faults in the area. These earthquakes are considered to be the most severe earthquakes associated with the faults and are assumed to occur at the point on the fault closest to the site. This defines the Maximum credible Earthquake and its location in terms of Richter Earthquake M and distance d to the causative fault. Time-history analyses of a dam are sometimes desirable. The required accelerograms may be produced by appropriate adjustment of existing or artificially generated accelerograms. The previously mentioned parameters are necessary considerations in the development of synthetic accelerograms or in the adjustment of actual recorded accelerograms." (United States Department of the Interior, Bureau of Reclamation, 1976)

(UNQUOTE)

This is exactly what the Project Authorities have actually done.

The Government of Gujarat had entrusted the Department of Earthquake Engineering, University of Roorkee with the task of carrying out studies of the seismotectonic setup of the region around SSP, static and dynamic analysis of the Navagam Dam and other related studies.

(QUOTE)

The Preface of the report at Annex IV (EQ 81-1, Revised) in January 1982 is reproduced below:

"The Government of Gujarat requested the Department of Earthquake Engineering, University of Roorkee to carry out studies of the seismotectonic setup of the region around Sardar Sarovar Project, static and dynamic analysis of the Navagam Dam and other related studies. A report entitled "Design Earthquake Parameters for Navagam Dam site in Gujarat (EQ 81-1)" incorporating the recommendations for design ground motion to be adopted for dynamic analysis of structures was submitted to the Project authorities in January 1981.

The Dam Review Panel of the Sardar Sarovar Project appointed by the Government of Gujarat reviewed the above report and has suggested that Maximum Credible Earthquake (MCE) and Design Basis Earthquake (DBE) be evaluated for the dam. The criteria to be applied with respect to the DBE is that no damage to the structure should result but permitting tensile stresses up to one eighth of ultimate compressive strength of concrete. For the purpose of defining the DBE, it was recommended that it be taken as half the intensity of the MCE. Two different inputs were recommended to be applied for the dynamic earthquake response analysis of structures; (i) Koyna accelerogram of December 11, 1967 adjusted to the intensity of the DBE (both the larger horizontal and vertical components of earthquakes to be applied simultaneously) and (ii) an artificial earthquake compatible with appropriate design response spectrum (Koyna earthquake spectrum smoothed and adjusted for geological difference at the Navagam dam site as compared with Koyna with intensity of the artificial earthquake defined by DBE). This report gives the recommendations for the MCE and the DBE and the time history of ground motion for the dynamic analysis." (Department of Earthquake Engineering, 1982)

(UNQUOTE)

The data on earthquake occurrence in Peninsular India show that the MCE may be of magnitude 6.5 and the seismotectonic setup of the region suggest that the MCE may occur along any of the capable faults/tectonic lineaments in this region.

Seismic risk at the Navagam Dam site was evaluated following the probabilistic approach taking into consideration the data on earthquake occurrence from 1917 onwards in the whole of the Indian-Subcontinent. The risk analysis was carried out by the Department of Earthquake Engineering, University of Roorkee.

The 1967 Koyna earthquake is the only earthquake for which a recorded accelerogram is available in India. The accelerogram recorded at the foundation gallery of abutment of Koyna Dam and reported by Krishna *et al.* (Krishna, Jai 1976) provides a time history of strong motion in the pre-Cambrian Continental Shield region covered with thick massive basalt flows estimated to be more than a kilometre in thickness at the Koyna dam site. The time history of strong ground motion to be adopted for dynamic analysis of Navagam Dam and appurtenant structures could be similar to that recorded during the 1967 Koyna earthquake, modified to the geological conditions of the area around Sardar Sarovar Project. Thus for the design of SSD, the accelerogram of the Koyna dam with appropriate adjustment has been considered.

With SSP poised for financial assistance from the World Bank, the Government of Gujarat, under advice by the Government of India, constituted a Dam Design Review Panel (DDRP) for review of the design of SSD, through a Government Resolution—Government of Gujarat, Irrigation Department, G.R. No. NMD/1881/56 K-4 dated 8-6-1981. The DDRP, consisting of eminent experts from India and abroad, examined the structural design of SSD and also reviewed the geology, hydrology and structural analysis of the main dam and its ancillary works, seismic potential, RIS potential etc. After going through all these aspects and also consulting other experts, the DDRP approved the design.

After the review and finalisation of the designs by the Panel, the construction of the dam, which had started with ancillary works in 1979, picked up tempo in 1988. However, at the time of writing the present paper, the work which had already progressed with the spillway portion of the dam up to elevation 80.3 m (about 65.0 m above the riverbed) has been at a standstill at this level since 1994.

The reservoir behind the dam has not led to any increased seismic activity, which is being monitored by the network of nine project seismological observatories already mentioned. The locations of these observatories around the reservoir were selected by IMD, Government of India. The type and number of seismic instruments at each of the observatories were finalised in consultation with IMD. The Jabalpur Earthquake was also recorded by the SSP network of seismological observatories.

After approving SSP for financial assistance, the World Bank suggested that a Dam Safety Panel comprising three independent experts be appointed to conduct periodic independent reviews during final engineering and construction of the dam and power houses. The Government of Gujarat accordingly constituted the Dam Safety Panel (DSP) with effect from 1.3.1985.

Studies related to aseismic design of Sardar Sarovar Dam, as also discussed earlier, were critically examined by the erstwhile DDRP. After detailed deliberations on the views given by various experts, the DDRP finally approved the design of the dam.

The Jabalpur Earthquake was considered by the petitioners apparently as a warning which "if ignored, (could wreak) devastation and destruction in the valley…" In this context, after the Jabalpur earthquake the DSP went into details of all relevant aspects from 9th to 12th July, 1997 and reviewed all the engineering provisions related to the SSD and the reservoir. The DSP in their report (Report of the thirty fourth meeting of Dam Safety Panel, Sardar Sarovar (Narmada) Project, Gujarat, July 9th to 12th, 1997, Kevadia/New Delhi) given to SSNNL, GoG concluded and recommended as follows:

> (QUOTE)
> "The Panel has made a comprehensive review of the Design Factors and design provisions for Sardar Sarovar Dam with particular reference to seismic safety. While doing so, the Panel has not only considered the up-to-date data from the seismic network established by the Project Organization in the Project area but has also taken due cognizance of the available recent additional seismic data and information regarding the seismic vulnerability of the Indian Peninsula which until recently was considered to be an a seismic

formation. As result of this review the Panel has concluded as following:

(i) Given the magnitude of the Project and its enormous benefits to the people of the participating states, the conservative approach adopted by the Project Organization in association and consultation with the established and dedicated Institutions as well as the well-known Indian and International Experts in field of seismology is affirmed as appropriate. The seismic design of Sardar Sarovar Dam developed with the State-of-the-Art technology ensures the capability of the Dam to sustain without failure even those seismic incidence(s) which have rather remote probability of occurrence.

(ii) The Pane has been observing all along the high standard of quality control and its enforcement by the Project Organization on the construction of the Dam works. This gives added confidence and assurance for satisfactory performance of the Dam. Our stipulation is that the performance of the Dam under severe earthquake incidence, whenever it occurs would not impair the Dam.

(iii) The recent Jabalpur earthquake is within the probable range of events likely to occur in the geotectonic set-up of Narmada area. It is covered fully by the parameters already provided for in the seismic design of the Sardar Sarovar Dam. No additional design measures of any kind are called for on account of this incident.

(iv) Panel suggests that the observations on the seismic network established around the Project area be continued to be observed and the data thus obtained be collected and suitably compiled concurrently. It would be advisable that a memo in future in this respect is presented for review by the Panel on an annual basis.

Review of the seismic data collected at the eight observatories of the Project Seismic Network indicates normal seismic activity which can be characteristically expected so far during 22 years of monitoring. The present network has also been serving a very useful purpose of constantly developing a desirable degree of awareness in the Project Organization about the probable seismic eventualities and related safety of the Sardar Sarovar Dam.

Project organization has installed a network of seismic observations around the reservoir area to obtain pre and post-impounding seismic events and their variation with time. Panel considers this to be a very useful provision which will be able to furnish regular data about seismic events before, during and after the creation of the reservoir. In the long term, such a network will be helpful in continuous monitoring of any significant seismic incidences especially form the strong motion recorders installed at the observation stations of the network." (Dam Safety Panel 1997: 13,14)

(UNQUOTE)

In this context it is affirmed that

The magnitude of Jabalpur Earthquake (M = 6.0) is well within the seismic parameters evolved and approved by the erstwhile Dam Designs Review Panel. That is to say that since the dam section of the Sardar Sarovar Project has been designed to withstand a ground motion generated by an earthquake corresponding to magnitude 6.5 occurring at a distance of 12 km, at a focal depth of 10 km, the effects of an earthquake similar to the Jabalpur earthquake having magnitude of 6.0, even if at all occurring at a distance of 12 km from the dam site, would be well within the forces for which the dam section has been designed.

The influence of Jabalpur earthquake on Navagam Dam area would be insignificant.

Even accepting that an earthquake of the same intensity as the Jabalpur earthquake were to occur in the vicinity of the dam, it would be fully taken care of by the design provisions made as clarified above.

Earthquakes have been occurring in India from times immemorial leaving their tell-tales in the form of shear zones, faults and other tectonic features. Some regions get big shocks, other moderate ones and some have no known history of their occurrence. Some areas get them frequently and others at long intervals of time. No location however could be described as "not susceptible" to earthquake occurrence.

Extract from the catalogue "Earthquakes of India and

Neighbourhood—From Historical Period up to 1979", published by the Indian Society of Earthquake Technology, Roorkee, 1983 gives an account of earthquakes with magnitude 5.0 and above which have occurred in India and near its International borders (between Lat. 48 Deg. N upto 11 Deg. N and Long. 62.7 Deg. E upto 99.6 Deg. E).

It may be seen that over 625 earthquakes of magnitude 5 and above have occurred in this region up to 1979. All these severe earthquakes have had no virtual ill effects on any of the major dams constructed in the seismically active Himalayan region. The Sardar Sarovar Dam is designed very conservatively with respect to seismic susceptibility of the region.

Issue No. III

Regarding: There is a possibility of the seismic activity being triggered off by the dams themselves—the scientific term being 'Reservoir Induced Seismicity' or RIS. They have quoted the Koyna Dam as an example for the largest RIS in the world and have also referred to Sukta Dam (Narmada Valley) and Bargi Dam being the cause for RIS (Jabalpur earthquake).

The phenomenon of RIS has been recognised ever since earthquakes (to magnitude 5) occurred in 1936, with 80 per cent the filling of Lake Mead (Hoover Dam on the River Colorado, completed in 1936 and until today the World's highest concrete dam, 221 m above lowest foundation), one of the largest man-made reservoirs of about 34,850 MCM capacity. Engineers and geoscientists have been concerned about whether a deep impoundment of water can induce macro-seismicity at a geologically susceptible site. Although there is no universally accepted proof that this can occur, after large number of site studies and analysis, circumstantial evidence of RIS exists only for some major reservoirs in the World.

A paper on Reservoirs and Earthquakes, by Ronald B. Meade, Department of Civil Engineering, Virginia Military Institute, Lexington, Virginia, United States of America, published in *Engineering Geology*, an International Journal, Vol. 30, No. 3/4, June 1991, The Netherlands, brings out the fact that in 1982 the US Army Corps of Engineers published a study of Reservoir Induced Seismicity. It is also mentioned in the said research paper that around 1981 the

US Army Corps of Engineers sponsored a study to critically evaluate the evidence of RIS. This study took an in-depth look after all the cases of damaging earthquakes that had been linked to reservoir activity in published reports prior to 1980. It thus evaluated the evidence of induced earthquakes from eight cases including Koyna, Hoover, Nurek and Kariba. Some relevant extracts and a listing of the eight cases of damaging earthquakes in Table below of the Research Paper are reproduced herein below:

> (QUOTE)
> "The practice of suspending judgment until the evidence can be evaluated is common to most scientific investigation. On the contrary, studies of reservoir induced Seismicity has routinely assumed a cause and effect relationship between the reservoir operation and earthquakes. In such a case the details are not evaluated but instead, form the basis for the characterization of reservoir induced Seismicity.
> The idea that stress changes due to reservoirs could trigger earthquakes is plausible. Reservoirs impose stresses of significant magnitudes on crustal rocks at depths rarely equaled by any other man-made work. The ability to prove a cause and effect relationship between reservoir activity and earthquakes is beyond the capabilities of scientists and engineers because of the severely limited ability to measure stress below depths of several kilometres.
> Despite the lack of direct evidence for evaluating a cause and effect relationship, the question of the role of reservoir activity in triggering earthquakes is sufficiently important to justify the serious review and evaluation of the available circumstantial evidence. About 1981, the US Army Corps of Engineers sponsored a study to evaluate the evidence of reservoir induced earthquakes critically (Meade, 1982). This study took an in-depth look at all the cases of damaging earthquakes that had been linked to reservoir activity in published reports prior to 1981. A listing of these cases appears as below.

Cases Reviewed in the 1982 Study

Reservoir	*Location Earthquake*	*Largest*
Hoover (Lake Mead)	USA	ML = 5.0
Kariba	Zambia	Mb = 5.8
Kremasta	Greece	Ms = 6.3
Koyna	India	Ms = 6.5
Kurobe	Japan	Ms = 4.9
Manic 3	Canada	Mb = 4.3
Hsinfengkiang	China	Ms = 6.1
Nurek	USSR	Ms = 4.5

ML = a local magnitude scale used.
Ms = magnitude measured from amplitude of surface waves.
Mb = magnitude measured from amplitude of body waves.

> The study used reservoir water level data and earthquake catalog data to judge the correlation between reservoir-induced stress changes and the occurrence of earthquakes. The correlation was measured by examining the number of earthquakes occurring during certain water level stages." (Meade, 1991: 245, 246) (UNQUOTE)

The author of the paper has also mentioned that the methods used in the 1982 study are applied to more recent data from Koyna and Aswan that appeared after publication of the said study in 1982. Some more relevant extracts from the said research paper are reproduced herein below:

> "Micro-earthquakes release so little energy that changes in the water level could account for all the energy released. Micro-earthquakes found near plutons, deep mines and volcanoes are examples of Seismicity generated by local stress changes. These small earthquakes are not related to large tectonically spawned earthquakes. In contrast to small shocks, the energy released by damaging earthquakes is so large that the stress changes caused by the reservoir are minor. Although the reservoir could possibly trigger a large shock, the energy changes caused by the reservoir should not increase the size of the maximum earthquake that could take place in a fault.

> The question of increased earthquake activity after impoundment is difficult to answer for areas normally considered earthquake country such as Greece, Japan, and Central Asia. In seismically active areas, the earthquakes occurring in the reservoir area may be unrelated to the reservoir. None of the literature summarizing reservoir-induced seismicity has attempted to separate coincidental earthquake activity from induced activity. These studies, that listed all reported cases of induced seismicity, assumed that all earthquakes near reservoirs were induced. This assumption is unwarranted and misleading." (Meade, 1991: 246-247)

Some additional results associated to Koyna earthquake (India) from the above mentioned research paper by Meade was—

> "In the years 1975 through 1980, 27 earthquakes of magnitude 4 or larger have occurred. Of these 27, eight occurred on September 20, 1980. These eight, plus three others took place in August and September during or immediately after a period of summer filling. The one other earthquake of 1980 occurred during declining water level. Of the 15 earthquakes that occurred from 1975 through 1979, five occurred during or immediately after summer filling and 10 occurred during the slow water level decline that occurs during most of the year. These data are summarised in Table below.
>
> "The recent data support the conclusion of the original evaluation that the Koyna seismicity was unrelated to the reservoir". (Meade, 1991: 257, 259)

Koyna earthquake data 1975-1980

Total number of earthquakes ML > = 4	27
Number occurring in 1980	12
Number occurring on September 20, 1980	8
Number during 1975 through 1979	15
Number occurring 1975 through 1979 during stability state N	10
Number occurring 1975 through 1979 during stability state T	5

Source: Meade, 1991.

> (QUOTE)
> Koyna and Aswan have been put forth as cases of induced seismicity. The evidence of induced seismicity is lacking in each of these examples. Use of these cases as example of induced seismicity has been built on the false assumption that every earthquake occurring on or near a reservoir is induced by the reservoir.
> It is evident that Energy changes caused by the reservoir will not increase the magnitude of the maximum earthquake that could take place on a fault. This immediately reduces RIS to a secondary-level phenomenon as compared to the natural earthquakes for which the dam has already been designed.
> – The tendency to ascribe all earthquakes in the vicinity of a reservoir to RIS is not correct, and can be misleading.
> – The Koyna Earthquake, which has been quoted as a specific case of RIS by the petitioners, is not actually so."
> (Meade 1991: 261) (UNQUOTE)

Recognising the moderate seismic activity in the Narmada Valley, the Government of Gujarat and the Dam authorities have installed a network of seismological observatories in consultation with IMD, Government of India, around the Reservoir periphery of the Sardar Sarovar Project to monitor the seismic status of the reservoir area, before, during and after impounding of the reservoir. These seismological observatories are equipped with state-of-the-art seismic instrumentation. In this connection, at the Central Data Processing Centre installed near the dam site, a sophisticated computer work station (DAC 300) has been installed which analyses the seismic data recorded by the network. Regular round-the-clock monitoring and collation of these data is done by the SSD authorities besides the Gujarat Engineering Research Institute.

Further, the contention that Sukta Dam has also triggered RIS is not made on any scientific basis or instrumentation monitoring. It would be rather unscientific to rely basically on either newspaper reports or alternatively records gathered based on "say of some vigilant villagers". All the data of micro-seismic activities have necessarily to be collected over a long period of time to assess the natural activity in the area and thereafter observe any changes in this activity as a result of creation of the reservoir in the back ground of essential/ significant criteria, and then analyse the result to arrive at any findings related to RIS. In this connection, it has been observed that Sukta Dam is only 38.12 m high, and its gross storage capacity is 89.29 MCM at FRL and is located about 400 km south-west of Jabalpur (shortest distance).

Extracts from the review of the discussions on Question No. 60, Reservoirs and the Environment—Experience in Management and Monitoring" during the Sixteenth International Congress on Large Dams held in Sam Francisco, USA in 1988, published in September 1988 issue of the *International Water Power and Dam Construction* Vol. 40, No. 9 are reproduced below:

> (QUOTE)
> "C.B. Viotti of Brazil then gave an account of a seismic monitoring programmed at ten reservoirs in south-eastern Brazil. The area, he said, was one of low natural seismicity, and the projects described had been monitored at various stages to see if seismic activity was being induced." (International Water Power and Dam Construction 1988: pp. 42)
> "He added that, in the past decade, there had been fewer cases of concern, possibly because of slower rates of impoundment now being implemented. He concluded that, in any case, early hazard identification would be particularly beneficial for new reservoirs in non-seismic areas, or when the increased occurrence of seismic events in already active areas had to be accounted for." (International Water Power and Dam Construction 1988: 43).
>
> (UNQUOTE)

It would be seen from the above that monitoring before, during and after impoundment of a reservoir is essential and important for assessing RIS and that a slow rate of impounding is helpful. In the case of Sardar Sarovar, seismic monitoring was started in 1970. As regards the benefits of slow impounding, the circumstances have already compelled its being done in stages.

The note on the Joint Inspection of Bargi Dam by officers of Central Water Commission and NVDA on 24.5.1997 states that the Bargi Dam, though 10 km from the epicentre of the Jabalpur earthquake, has not suffered any damage.

Thus it would be quite evident and clear from the above discussions that any concerns about the safety of SSP and the main dam, as far as seismicity is concerned, are unfounded. Accordingly any apprehensions regarding the safety and vulnerability of the main dam due to seismicity do not now find any place here, as the Dam is safe.

5

Siltation Study of Sardar Sarovar Project with Present Status

B.J. Parmar and *K.G. Rathod*

1. Introduction

The Narmada, the longest west-flowing river of peninsular India and 5th longest in India, rises near Amarkantak in the Anuppur District (carved out of Shahdol district in August 2003) of Madhya Pradesh in the Maikala range, at an elevation of about 900 m. The total length of this inter-state river from the source to its outfall into the Arabian Sea (Gulf of Khambhat) is 1312 km, of which the first 1077 km is in Madhya Pradesh. In the next 35 km, it forms the boundary between Madhya Pradesh and Maharashtra, and in the following 39 km between Maharashtra and Gujarat. The last 161 km lies in Gujarat. Of its 41 tributaries, 22 are on the left bank and 19 on the right bank. The Narmada basin extends over an area of 98,796 km^2. It is bounded on the north by the Vindhyas, on the east by the Maikala range, on the south by Satpuras and on the west by the Arabian Sea (see map, Fig. 1).

It is proposed to build 31 major dams, 135 medium dams and 3000 minor schemes during various stages of the integrated development of the river valley. Of the major dams, Sardar Sarovar (SS) with a length of 1210 m and 167 m maximum height, built in 7 million m^3 of concrete, is the terminal dam. As of now the work on it has been completed up to crest level and the radial gates are yet to

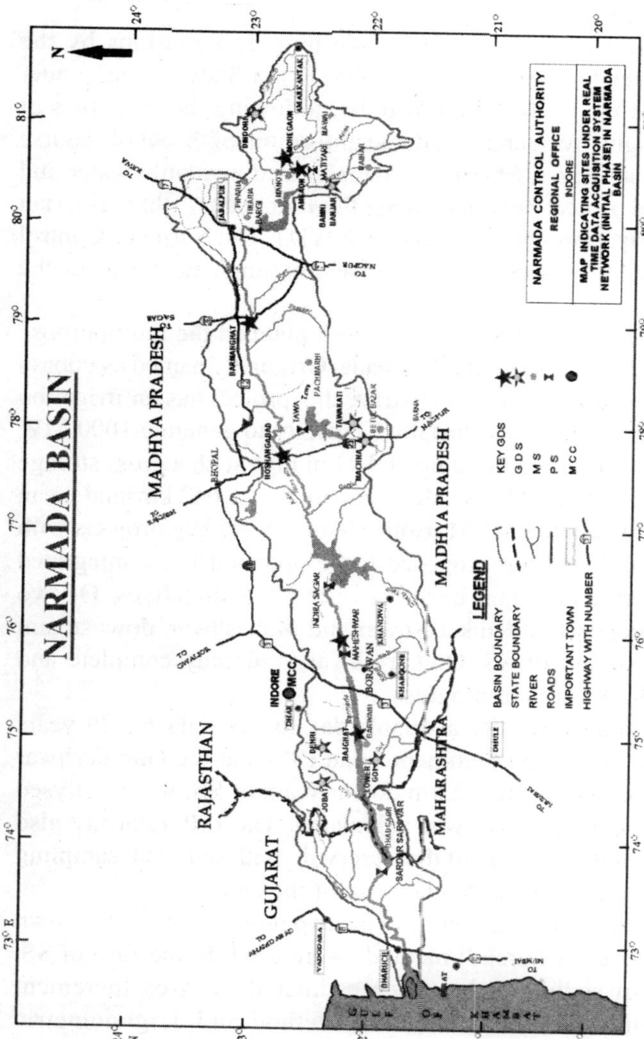

Figure 1: Map of the Narmada Basin

be erected. The reservoir has vast storage capacity of 9500 MCM (7.7 MAF) with live storage capacity of 5800 MCM (4.75 MAF). Its catchment area is 88,000 km^2 and the mean annual runoff at the dam is 33,600 MCM.

The project emerged from decade-long deliberations by the Narmada Water Disputes Tribunal (NWDT), a body formed under Article 262 of the Constitution of India, following the principles of 'equality of right' over, and 'equitable utilisation' of the whole course of an inter-state River. Sharing of the assessed available water and the associated project benefits among the four states—three riparian and one non-riparian—was decided by NWDT. The Narmada Control Authority (NCA) was created to oversee the implementation of the project.

The Government of Madhya Pradesh planned the multipurpose Indira Sagar (IS) project on the Narmada. Originally named as Punasa Dam and located in Khandwa District, this project has an irrigation potential of 0.265 Mha and installed capacity to generate 1000 MW of power. The dam is 653 m long and 92 m high with a gross storage capacity of 12,220 MCM. Its catchment area is 61,642 km^2 and mean annual runoff is 26,465 MCM. Both SS and IS are key projects with sizeable storage, and are proposed to be operated in an integrated manner to derive optimum benefit as per NCA directives. Of two other dams proposed at Omkareshwar and Maheshwar, downstream of IS and upstream of SS. Omkareshwar is already complete and work on Maheshwar is in progress.

For silt studies, readily available day-to-day data for 29 years (1962-90) of Suspended Sediment Load (SSL) at the Garudeshwar Gauge and Discharge site, 12 km downstream of SS, were analysed and average sediment load was calculated. The ISP authority also estimated the sediment load of that reservoir, with sediment sampling at Mortakka, about 60 km downstream of the dam.

To arrive at the trap efficiency of the two projects, the well-known Brune method and Churchill methods were used. In the case of SS, distribution of sediments has been estimated by Area Increment Method, Empirical Area Reduction Method and Trigonometric Method. ISP intercepts about 70 per cent of the catchment area and 77.67 per cent of average annual runoff of the SS dam. IS would, therefore, trap considerable amounts of sediment. When the SS works were actually taken up and IS was still to commence (although both were planned to be simultaneously built and operated) the expected life of SS was assessed considering alternative dates for the completion of IS.

2. Estimation of Total Sediment Load Transported by the Narmada at Garudeshwar Gauge Site

2.1 Suspended Sedimentation

As mentioned above, studies for the SS Project were based on daily data of suspended sediment in the river at Garudeshwar, for 29 years from 1962 to 1990. The water samples were collected with a bottle type sampler. Suspended sediment transported by river flow was analysed to measure the distribution and concentration of sediments by particle size category, defined as Coarse Suspended (> 0.152 mm), Medium (0.07-0.152 mm) and Fine (< 0.07 mm) sediment. The concentration of these three categories of sediment was calculated in gm/l.

The average volume of suspended sediment transported by the river at Garudeshwar is an estimated 4.10 ha-m/100 km^2/year, or 1000 ppm. Assessing the progressive sedimentation in the reservoir and its useful service life requires long-term runoff data. Partly observed and partly hind cast runoff data of the Narmada are available for the period 1891-2009. The average annual runoff from these data is computed at 40,480 MCM. The average runoff for the short-term (1962-1990) observations is 36,110 MCM. From this it can be deduced that SSL varies proportionately with runoff, and the rate of SSL carried by the river computed from 29 year data, is increased in the ratio of average runoff for 1891-2009 and that for 1962-1990. Considering runoff series from 1891 to 2009, the SSL works out to 4.596 ha-m/100 km^2/year.

Observed data of runoff and SSL in the river at Mortakka Gauge Site are available for 1951-1952, 1956-1959 and 1962. The average annual runoff and average annual sediment load for these seven years are computed respectively at 26,440 MCM and 4.288 ha-m/100 km^2/year. With hindcast series for runoff at the IS site for 1891-2008, these averages work out respectively at 34,930 MCM and 5.665 ha-m/100 km^2/year.

2.2 Bed Load with Total Silt/Sedimentation Load

Observations for measuring sediment load transported by the Narmada along its bed have not been done at Garudeshwar. It has not been possible so far to collect samples of bed load from the river during the monsoon—the experience for all rivers in India. The bed load for Indian rivers is generally estimated using a correlation between SSL and bed load developed by the United States Bureau of Reclamation (USBR) for rivers in USA. The relevant parameters for this correlation are SSL, type of river bed material and texture of suspended sediments for the river for which bed load is to be assessed.

As mentioned above, the average volume of SSL in the Narmada at Garudeshwar is about 1000 ppm. The river bed here, and for a long stretch upstream, is either rocky or covered with deposits of sand, gravel, cobble and boulders. Suspended sediment material *inter alia* contains about 25 per cent sand. According to USBR findings, for a river with a rocky or gravely bed, carrying a SSL of 1000-7500 ppm, the proportion of bed load is in the range of 5-12 per cent of SSL.

Many researchers have tried to correlate bed load as a fraction of SSL. However, very large variations in results have been found, because they are either based on theoretical formulae or model studies in which particles are of silt size. Actual field measurement for assessment of bed load is difficult and, therefore, it is also difficult to establish a correlation of bed load and SSL which is actually collected by sampling. The task becomes more difficult with large rivers like the Narmada, where it is not easy to collect samples either with bed load samplers or by trenches in the river bed.

The above constraints led the project authority to seek collaboration of Gujarat Engineering Research Institute, Vadodara to evolve a method to measure bed load. In the research and development for this during the years 1979-1989, GERI conducted experiments at Tilakwada, about 25 km downstream of the dam, using a new approach to assess the nature and quantum of bed load. Large pits were dug in the sand/gravel shoal of the river and bed load was determined by correlating the flow over the pit with the entrapped quantities of bed load in such pits.

July-September is the annual flood period in the Narmada. Trial pits were dug at various locations selected with due consideration of low to high level river bed, so as to trap bed material in pits during floods of low to high magnitude. For the first experiment several pits of size 1.5 × 3 × 3 m, with upstream and downstream slopes of 1:5 and 1:1 respectively, were excavated in May-June 1979 at selected locations in the shoal. Their beds and sides were lined with a 20 mm thick hand dash *in-situ* coat of cement mortar (1: 6) to impart stability to slopes and to prevent mixing of bed load with in-situ material. The pits were carefully aligned parallel to flow direction to match the visually observed flow patterns.

Initially, owing to their small size, the pits did not serve their purpose of measuring the rate of transport of bed load. In later years (1982-1988) experiments were conducted with different sizes of pit to obtain results with partial filling. This revealed that increased depth played an important role and the location of pits from lower to higher elevation, changed to match the flow pattern, also helped in the required observations. In 1988 three pits at lower, average and higher elevations, and sizes as large as 7 × 30 × 5 m were tried in the same cross-section to reduce flow width for trapping bed load to get partial filling.

During the experiment period up to 1988 all the pits were fully filled up, indicating that larger sized pits were required to obtain the rate of bed load. Results for the ten-year period indicated that with respect to the average SSL of 1000 ppm at Garudeshwar without the dam, bed load range as percentage of SSL was in the range of 1.2-3.0 per cent.

2.3 Total Siltation/Sedimentation Load

As indicated at 2.1 above, the average volume of SSL transported by the river at Garudeshwar site is estimated as 4.10 ha-m/100 km^2/year from the observed data of 29 years (1962-1990: pre-dam condition). Adjusted for average runoff for the period 1891-2009, the SSL works out at 4.596 ha-m/100 km^2/year.

GERI's observations and the guidelines given by USBR suggest a bed load 5 per cent of measured SSL at Garudeshwar, which comes to 0.2298 ha-m/100km^2/year. Thus the average total sediment load would be 4.826 ha-m/100km^2/year. Considering bed load as 5 per cent of suspended load for the IS Project and increasing pro rata suspended sediment for the series 1891-2008, the total sediment load worked out as 5.948 ha-m/km^2/year.

3. Trap Efficiency

Trap efficiency is the ability of reservoir to trap and retain sediment. It is expressed as a percentage of sediment yield (incoming sediment) retained in the basin, with the rest passing on downstream. It depends *inter alia* on the sediment load characteristic, the detention time of inflow, the age of the reservoir and the kind of outlets provided in the dam for release of water. Sedimentation Index is defined as the ratio of period of retention to mean velocity through the reservoir. It is expressed as rates of deposited sediment to total sediment inflow. Brune and Churchill have evolved methods to evaluate the trap efficiency. In Brune's method, trap efficiency is related to inflow into the reservoir and its storage capacity. At SSP the average annual inflow during monsoon, considering the 2nd stage development of the Narmada basin, is computed as 15,270 MCM and its gross storage capacity is 9500 MCM. Thus the capacity-to-inflow ratio works out at 0.622. As per Brune's correlation curve the trap efficiency of SSP is 96 per cent. Churchill analysed data of the Tennessee Valley Reservoir and developed a relationship between percentage of incoming sediment passing through the reservoir and its sedimentation index. Using relevant data of SSP, the sedimentation index works out to 5.47×10^8. According to the correlation developed by Churchill, the percentage of sediment passing through SSP is found to be 2.3 per cent and trap efficiency is 97.7 per cent. All studies on SSP use a trap efficiency of 98 per cent.

In the case of IS, the gross storage capacity and average annual inflow are 12,220 and 34,930 MCM respectively. The capacity-to-inflow ratio is 0.349. Using Brune's correlation, trap efficiency works out to 93 per cent. Using data of reservoir capacity, inflow and reservoir length, the sedimentation index for IS is computed as 1.918×10^8. From Churchill's correlation, the proportion of sediment passing through IS is 4 per cent. This gives trap efficiency of 96 per cent, which is assumed for sedimentation in IS.

The amount of fine sediment entering SS from the catchment area above IS is calculated at 96.323 ha-m/year. The total annual sediment load from the intervening catchment area is 552.46 ha-m/year, of which the fine sediment load area is 430.80 ha-m/year. Thus, the gross annual fine sediment entering SS is 527.12 ha-m/year, and at a trap efficiency of 98 per cent the volume of fine sediment trapped and deposited in SS is 516.58 ha-m/year. The total annual sediment deposit here comprises coarse sediment from intervening catchment area, total fine sediment trapped and deposited, and bed load equivalent to 5 per cent of SSL. Coarse sediment from intervening catchment area is worked out at 121.66 ha-m/year, fine sediment is 516.58 ha-m/year and bed load is 27.62 ham/year. The annual rate of silt deposition in SS is thus 665.86 ha-m/year. This calculation assumes the simultaneous construction of SS and IS. If SS were built before IS, the average annual sediment entering SS during the time lag would be 4,246.80 ha-m/year and net amount of sediment deposited would be 4,161.77 ha-m/year. Since IS has been completed ahead of SS, this scenario is no longer relevant to this study.

4. Expected Service Life of SSP

4.1 As the issue of sedimentation was persistently raised by environmentalists and activists, the Narmada Planning Group (NPG) of the Government of Gujarat critically examined the problem in order to determine the feasible service life of the reservoir. With modifications based on the new rate of sediment worked out from the long-term inflow data of 1891-2009, Table 1 below gives the Feasible Service Time of SSP as estimated using the following three different methods and for different construction schedules of SSP and ISP:

1. Empirical Area Reduction Method
2. Area Increment Method
3. Trigonometric Method

Considering the actual developments to date, Case nos. II and III are now irrelevant.

Table 1: Feasible Service Time of SSP (in years)

Sl. No.	*Particulars*	*Feasible Service Time of SSP*		
		Empirical Area Reduction Method	*Area Increment Method*	*Trigonometric method*
Case I	Completion of SSP and ISP together	499	705	722
Case II	ISP completed 5 years after SSP	473	679	696
Case III	ISP Completed 10 years after SSP	446	653	670

5. Conclusions

5.1 From Table 1 it is evident that maximum life of the SSP can be anywhere from 499 to 722 years if the completion of SSP and ISP is synchronous. In fact ISP has been completed ahead of SSP.

5.2 About thirty major projects are either existing or proposed on the Narmada and its tributaries. Also there are several medium and minor projects planned. Therefore, it is expected that upon completion of all these major, medium and minor projects in the basin, their combined trapping efficiency for suspended sediment and bed load would considerably reduce the volume of sediments and bed load entering the terminal SSP. In effect, the life of SSP would be augmented considerably beyond the period estimated in the above table.

5.3 The total catchment area of the Narmada at the SSP site is 88,000 km^2. In 1983 the Government of India constituted an Inter-Department Committee to study the existing situation in the catchment area and to suggest necessary measures to minimise soil erosion. Detailed surveys and studies were carried out as proposed in the environmental clearance prior to the preparation of an action plan for the catchment area. The catchment area treatment (CAT) in forest and non-forest areas is being done at the same pace as the stages of construction of the dam.

The work of measuring effectiveness of these treatments was assigned to the Central Soil and Water Conservation Research and

Training Institute (CSWCR&TI), Vasad in September 2002. The Institute, in consultation with the Forest Department, identified and established six monitoring stations for data collection during monsoon in representative micro-watersheds on both banks of the Narmada Catchment across three forest divisions. They collected data of rainfall, runoff and sediment during the monsoon for the years 2004-2008. Analysis of these data reveals that the rate of silting in most of the watershed has been less than 5.34 ha-m./100 km^2/year.

In addition, the assessment of forest cover using remote sensing technology was done by Gujarat Ecological Education and Research (GEER) Foundation, Gandhinagar. The satellite imagery of 1986, prior to the implementation of CAT, and that of 2006 shows a remarkable increase in vegetation cover this can be seen from chart below:

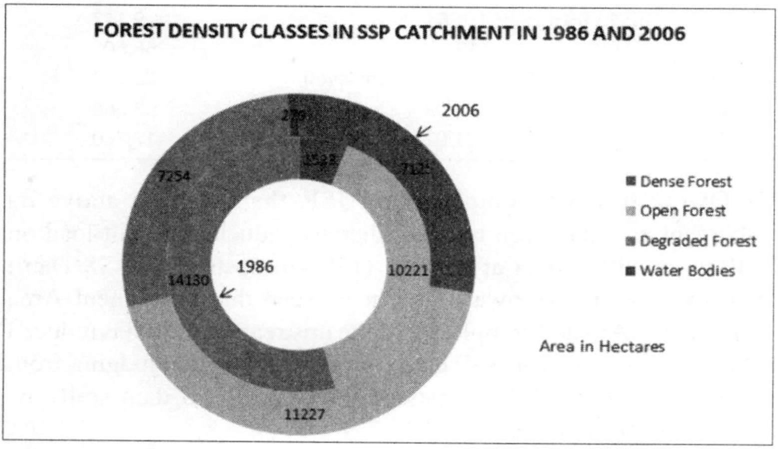

The GEER Foundation study reveals that the Dense Forest cover has increased from 1,533 ha (5.9%) in 1986 to 7,125 ha (27.52%) in 2006—an increase of 21.62 per cent in 20 years under this class. Most of the area under Open Forest got transferred to Dense Forest. The former has increased from 10,221 ha (39.48%) in 1986 to 11,227 ha (43.37%) in 2006—an increase of 3.89 per cent. The area under degraded forest reduced from 14,130 ha (54.6%) in 1986 to 7,253 ha (28.03%) in 2006, reflecting a 26.57 per cent reduction in area under this class.

Construction of SS dam was started in the year 1987 and, proceeding in stages, reached EL 69.0 (minimum block level) in 1994. The construction level achieved at the end of June for respective years for IS Project (ISP) and SS Project (SSP) are given in Table 2 below:

Table 2: Construction level achieved for ISP and SSP year-wise

Year	*ISP (level in m)*	*SSP (level in m)*	*Max Computed flood at SSP (cumec)*
1999	208	85	30,303
2000	208	85	4,644
2001	208	90	9,657
2002	215	95	28,802
2003	225	100	13,962
2004	246.13	110.64 (MDDL)	13,593
2005	262.13 (FRL)	110.64	9,147
2006		119	56,500
2007		121.92 (Crest level)	22,798
2008		121.92	3,568
2009		121.92	17,700

Owing to early completion of ISP, the silt from above its catchment area has been trapped, thereby reducing the silt load on SSP. From silting data at Rajghat (130 km upstream of SS Dam) from the year 1994 onward, it can be seen that Catchment Area Treatment (CAT) and completion of the upstream project has reduced silt rate in SS dam. The SSP dam is not complete, and judging from Table 2, major floods have passed through the SS dam spillway, flushing silt from the dam.

5.4 The Forest Department has prohibited cutting of trees on the periphery of the reservoir. This is a positive measure helping to control the flushing of sediments into the reservoir. Besides this, in SSP and ISP, enough silt traps have been provided.

5.5 Correct assessment of sedimentation rate is of great concern as it has a bearing on all aspects of the utility of a reservoir. In the past, silt measurement involved the use of equipment such as theodolite, levelling instrument, sextant, range finder, sounding rod, echo sounder, FRP boat etc. This was not only time-consuming; it also presented problems of accuracy in wide reservoirs of large dams like SSP. Moreover, the measurement grid was normally 1 km. The modern technique, procured by GERI, uses a boat mounted with an integrated bathymetric system for hydrographic survey. It employs Total Station or Remote Sensing data for computing the capacity of the exposed portion of reservoir up to FRL. Satellite imagery is used for measuring spread area at different water levels and reservoir capacity is calculated. When the gates are installed on the spillway at SSR, it is planned to use this technique for sedimentation survey of the reservoir with a closer grid. From the results so obtained, it will be possible to know the actual sedimentation against the calculated rate of siltation/sedimentation.

5.6 When the reservoirs are in cascades on a river such as the Narmada, their operation during high floods also needs high precision. It implies impacts related to instability of boulders on adjoining hill, especially in narrow reaches of the river flowing through narrow valleys, resulting in heavy landslides. In the case of SSP this is not expected as the terrain is quite stable (see Java G.L., Chapter 4 of this volume).

5.7 Efficient sedimentation management sometimes needs passing the sediment downstream of the reservoir by constructing low diversion dams upstream and diverting them through a channel or tunnel capable of bypassing large floods carrying a bulk of bed load downstream of the main reservoir. From the details presented above, it can be concluded that this is not the case for Narmada Valley.

6

Final Stage Diversion of Narmada River Fair Weather Flows during Construction of Sardar Sarovar Dam

B.J. Parmar and *R.L. Gupta*

1. Introduction

The Narmada is the longest west-flowing river in peninsular India. Rising in Central India, it flows about 1300 km to the Arabian Sea (Fig. 1).

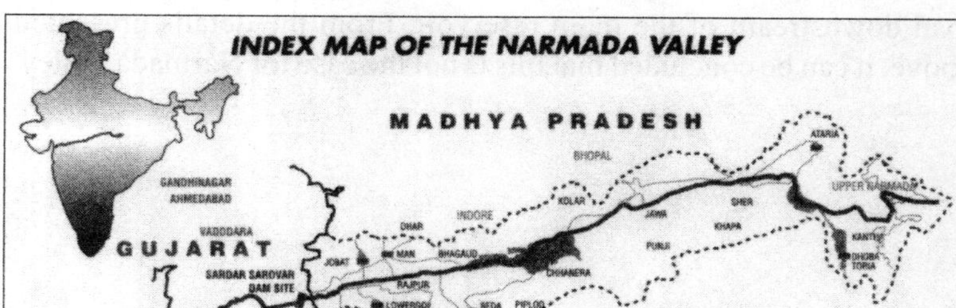

Figure 1: Location Map of Sardar Sarovar Dam Site

Four states of central and western India—Madhya Pradesh, Maharashtra, Gujarat and the non-riparian Rajasthan—will benefit from the 34,537.50 M.cu.m. of water from the Sardar Sarovar Dam, the terminal dam on the Narmada, now under construction. This unique project is planned to irrigate 1.905 Mha of land in Gujarat, Maharashtra and Rajasthan, generate 1450 MW hydro-power and provide drinking water to a population of 20 million in Gujarat.

The maximum design flood of the Narmada from the catchment area of 88,000 km^2 above the dam site is exceptionally high at 86,450 cumec. The maximum probable flood has been assessed at 169,800 cumec. The Sardar Sarovar Dam is a gravity mass concrete structure, 1210 m long, with an average height of 128 above the river bed, involving 6.82 million m^3 of pre-cooled concrete. The dam after completion will create a reservoir with a live storage of 5800 MCM (4.73 MAF) extending to about 214 km in the Narmada valley and submerging an area of 37,000 ha. Power generation is through an underground river-bed power house (1200 MW) and a canal head power house on the right bank (250 MW).

2. Original Diversion Scheme

At the commencement of construction, the high discharge of the river in the monsoon months allowed work from 15 October to 15 June, when the average flow ranged between 700 and 200 cumec. This changed with the commissioning, during 1990-91, of the Bargi Power House about 800 km upstream in Madhya Pradesh, contributing regulated releases all the year round.

The channel at the dam site has a rocky bed and is about 500 m wide. During the monsoon the river occupies the entire width of the channel. The maximum flood recorded (1970) is 69,385 cumec. A detailed study of the topographical and geological features of the site revealed that it was not possible to divert such a huge discharge through tunnels or by any other means to allow construction of the dam. Sardar Sarovar, being a concrete gravity dam for its entire length, presented no problem in allowing monsoon flow over its partly built blocks during the construction period. Therefore, it was planned to divert only the post-monsoon flow in stages during construction of the dam. This diversion and care of the river during construction was contemplated in three stages:

2.1 First Stage Diversion (Sill Level of Construction at RL 18.20 m, i.e. 0.20 m Above the Bed of the Diversion Channel)

In the first stage the river was diverted to facilitate treatment and subsequent grouting of the fault zone in the deepest portion of the channel. The flow was diverted through 10 construction sluices of size 2.15 m × 2.75 m in spillway block nos. 35-36, situated in the open cut diversion channel, keeping the sill level of construction at RL 18.20 m, i.e. 0.20 m above the bed of the diversion channel (Fig. 2).

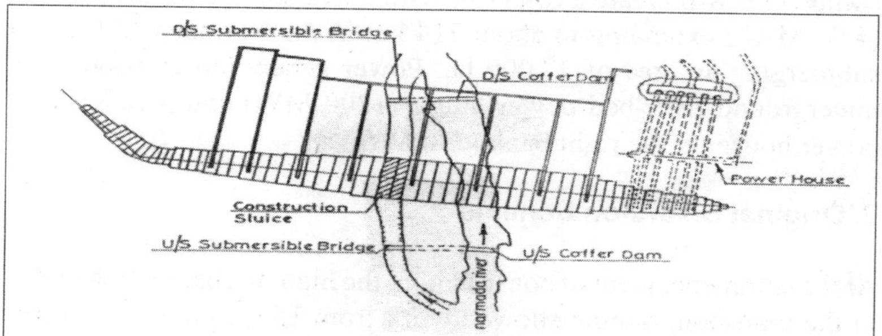

Figure 2: First Stage River Diversion Scheme

The following works were also carried out to ensure effective functioning of the first stage diversion:

(i) Semi-permanent coffer dams, one upstream and one downstream of the dam seat.
(ii) Open cut diversion channel to the left of the fair weather river flow channel.
(iii) Submersible bridges across the diversion channel, one each upstream and downstream of the dam seat.
(iv) Construction of two spillway blocks, nos. 35-36, up to RL 21.87 m, to accommodate 10 construction sluices.
(v) Excavation for spillway blocks foundation within safe blasting limits alongside blocks 35-36, constructed in the diversion channel portion.

2.2 Second Stage Diversion (Sill Level at RL 35 m)

In the second stage it was planned to re-divert the river to facilitate construction of sloping aprons in and alongside the diversion channel. To this end, it was planned to build another set of eight construction sluices in spillway blocks 44-45, with sill level at RL 35 m. Before initiating this diversion, all the blocks in the river gorge portion were required to be raised to RL 43 m. The first stage sluices were to be plugged soon after the second stage construction sluices became operational. This stage envisaged the following works:

(i) Extension of blocks 35-36 to accommodate construction sluices for the first stage river diversion scheme to suit the final section of the dam. This was done in May 1989 by closing the sluice opening with gates and diverting the river through block 38.
(ii) Raising the blocks in the river gorge portion to RL 43 m.
(iii) Excavating an open cut tail channel beyond the end sill of blocks 42-46.
(iv) Construction of eight sluices with sill RL 35 m in blocks 44-45.
(v) Construction of stilling basin for blocks 44-45.
(vi) Construction of upstream coffer dam for construction of block 37.
(vii) Construction of downstream temporary coffer dam beyond the downstream submersible bridge to keep the areas of the stilling basin for blocks 33-44 free of river water from downstream.
(viii) Lowering and subsequently re-covering the gates used for facilitating plugging of construction sluices in blocks 35-36.

2.3 Third Stage Diversion (Sill Level at RL 60m)

The third stage diversion of flow was originally planned through the permanent river sluices to be formed in the body of the dam blocks 44, 46, 48 and 50 with sill at RL 60 m. Simultaneously, the left non-overflow auxiliary spillway blocks and right bank power dam blocks were planned to be raised to a level to prevent overflow during the construction stage for a flood discharge of 69,385 cumec. In the river gorge portion the dam blocks were to be raised to a minimum RL of 69 m and the sluice blocks to RL 85 m to accommodate the hoist chamber of the service slide gate. It was planned to plug the eight construction sluices of the second stage when the flow could be let through the river sluices.

3. Revision of Diversion Scheme

The Government advanced the construction programme by a year. Accordingly the dam works were rescheduled for completion by January 1998. In the original programme water for irrigation and power generation was to be diverted after June 1995, for which the dam was to be raised to RL 85 m by June 1994 and to RL 112 m by June 1995. But the decision to accelerate construction meant that it had to be raised to RL 95 m by June 1994 in order to enable diversion of water for irrigation after June 1994. Thus, it emerged that the second stage construction sluices were required only for one season, leading to a review which resulted in omitting construction of the second stage construction sluices in blocks 44-45.

Finally, it was decided to divert the post-monsoon flow of the river through permanent sluices with sills at level 53 m in blocks 44, 46, 48 and 50 after closing the first stage construction sluices in blocks 35-36. This required the first stage diversion scheme through the ten construction sluices to remain operational until the permanent river sluices were ready and provided with gates to control post-monsoon flows. The final stage diversion was accordingly planned to be completed by March-April 1993 in order to meet various deadlines. This entailed the following additional works:

(i) Raising the first stage construction sluice piers of blocks 35-36 from the previous levels of 20.955 m/21.87 m to RL 35 m to enable lowering of gates for closing the sluice openings.

(ii) Raising all the NOF blocks and power dam blocks to minimum RL 93 m and the river gorge portion blocks to RL 69 m. Blocks with river sluices were to be raised to RL 85 m to accommodate the hoist chamber for operating service slide gates of river sluices, gate gallery etc.

(iii) Completing service spillway stilling basin nos. 4-5 along with right training wall to at least RL 40 m and dividing wall nos. 42 and 47 to their finished profile.
(iv) Excavating an open cut tail channel beyond end sill of bay nos. 4 and 5 and constructing a temporary submersible bridge across the tail channel.
(v) Installation of service slide gates for operation of river sluices.
(vi) Cleaning and maintaining the existing open cut diversion channel free of shoal deposits during the working seasons till final stage diversion.
(vii) Installing winches at RL 67 m and head gear arrangement at RL 35 m at blocks 35-36 to enable simultaneous lowering of gates in all sluices (Image 1).

Image 1: Head Gear Arrangement for Lowering of Sluice Gates

A detailed description of the aforesaid activities to be completed before final stage diversion is given below.

4. Raising First Stage Construction Sluice Piers

4.1 As mentioned above, the construction programme was advanced by a year. Thus the only option available was to use the first stage construction sluices at RL 18.20 m for the second stage diversion too. This required raising the piers of the sluices to RL 35 m with a vertical face, creating a working platform of about 5 m width at this level to accommodate winches for lowering of sluice gates. Thereafter the blocks were to be raised with the usual upstream batter of 0.4H:1V (Fig. 3).

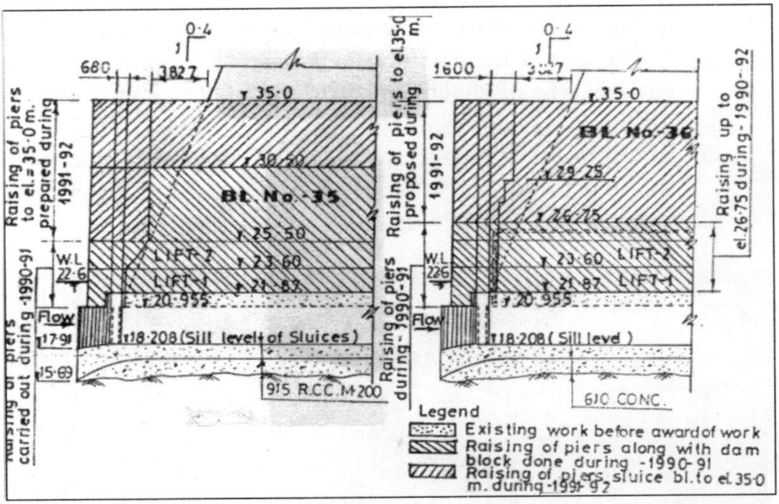

Figure 3: Proposed Raising of Sluice Piers of Block Nos. 35-36

4.2 During the month of May the water level was at around RL 25 m owing to the silting of upstream and downstream diversion channels and inflow of both the units from Bargi Power House and base flow of the catchment. The Government of Madhya Pradesh was persuaded to close one unit, bringing down the river to RL 22.60 m upstream of the sluices even after clearing of the diversion channel.

4.3 In the given situation, only two options were available: (1) to raise the piers under water flowing at 4-5 m/sec—a stupendous task involving high-precision and micro-level planning besides being risky owing to the limited time available for concreting at the end of the working season; or (2) to take up the work when closure of both Bargi units could be arranged. As the second option was very remote, it was decided to go ahead with raising the piers under water up to RL 25.50 m within a period of 3 weeks, when the Government of

Madhya Pradesh agreed to the closure of one Bargi unit. Raising the piers from RL 20.955 m to RL 25.50 m within three weeks was a challenge that called for an aggressive strategy:

(a) Work under flowing water was not possible. The flow had to be stilled on both side of the piers to be raised. This required closure of two sluices by lowering gates. A team of competent divers was deployed to work underwater to facilitate lowering of gates and steel shells for raising of piers aligned with the pier grooves (Image 2).

Image 2: Lowering of Steel Shells for Raising Pier Height

(b) Lowering of gates in the sluices required dummy steel frames at both ends at the top of the gate, the frame size being a little shorter than the pier groove dimensions. The dummies protruded above the water level so as to guide the lowering of the steel shell into its correct location under the heavy water current.

(c) Pier shells were made of 8 mm steel plate, stiffened and braced, placed in position and filled with concrete of suitable grade. Their height was fixed so that the first shell projected well above the water level when placed in position with the adjoining sluice openings closed. The designed height of this shell (liner type-I) was 2.6 m (RL 20.955 m to 23.60 m). The shell for the pier nose and the grooves for the first lift were fabricated as a single unit to enable divers to work within the shell and correctly align the grooves of existing piers with grooves of the shell.

(d) The shells for the second lift of the piers, from RL 23.60 m to 25.475 m were fabricated in 3 parts—pier nose in one part and gate grooves in two.

(e) Piers in both the dam blocks were raised simultaneously so as to complete the work within the stipulated period of three weeks. The upstream profile block no. 35 slopes in a curve from RL 23.6 m to 25.5 m. Two-tier platforms were made to facilitate work on this block.

(f) The grooves were cleared of all foreign matter to allow lowering of gates right down to sluice sill level. The corner angles in the existing concrete in the upstream face of pier grooves were found bent and were obstructing the lowering of gates. These and other obstructions were removed.

4.4 The work described above started on 21 May, 1991. Experienced engineers/supervisory staff, skilled works and divers, machinery with adequate reserves, were deployed to minimise delays. Raising the piers was complicated work and presented a lot of unforeseen problems. It required high precision and meticulous planning. About 40 tonnes of steel shell fabrication was done round the clock in two weeks. The shells for pier grooves, fabricated according to the theoretical size, had to be modified for each individual pier to match the existing faces of pier grooves.

4.5 Such precise work, to allow smooth lowering of the gates in the sluices, demanded perfect matching of thee grooves of steel shell with those of existing piers. Dummies were welded on the top of the sluice gates and then lowered in existing grooves for a perfect match between the piers and the shells. This was accomplished by competent and skilled divers.

Transporting the shells and lifting them onto the blocks with the mobile crane required careful handling so that the shells were in no way deformed. After lowering the sluice gates on both sides of the pier to be raised, the shell was placed with the dummy lowered in the pier groove.

The divers had assessed, before lowering of gates, the obstructions in the grooves; these were removed by underwater steel cutting.

After lowering the gates the steel shells were gradually lowered onto the existing concrete piers, guided by dummies. In some cases the shell did not match the shape of the groove in the existing pier. The divers noted the deviation, lifted the shell and returned it to the fabrication shop for appropriate modification and return to the site. At times the process had to be repeated to finally position and align the shells properly within 5 mm tolerance with the help of large turnbuckles.

The task was still not over. After alignment each shell was held in position by welding steel supports from the inside, so that it did not get displaced by water flow and/or concreting. Sand-filled gunny bags placed at the bottom of the shell reduced the large gaps due to irregular shape of the earlier concrete work damaged during 10 years of sluice construction. In some places, *in-situ* steel plates were fitted with nuts and bolts under water in gaps due to the damaged portions of existing concrete of the pier grooves. Precise engineering standards dictated the placing of perforated MS pipes inside the shells to anchor the old and the new concrete and grouting.

The $A_{75}S_{350}$ CL 450 kg/m^2 grade concrete for the piers was transported from the batch mixing plant in buckets, in lots of 3.0 m^3 each. This minimised pouring time and reduced the effort involved in washing fines and cement from poured concrete that might lodge in the likely gaps between the form and the older concrete. However, the sealing done by the divers was so perfect that the need for such washing was negligible (Image 3).

After concreting of the piers, the gates were lifted. On one occasion during this operation a pier got disturbed. Its shell, with concrete, had to be dismantled and discarded and the entire pier recast.

The work continued against heavy odds until 10 June, when a flood of about 3500 cumec reached the dam site, and the raising of the piers to RL 35 m could not be completed as planned. In block 35 they reached RL 25.475 m and in block 36 to RL 25.50 m. It was decided to continue work in the next season (1991-92) without closure of Bargi Power House.

5. Scheme of Lowering Gates for Closure of First Stage Sluice Opening

5.1 Blocks 35-36 were raised to RL 67 m by the end of February

1994. At this level winches of 18 tonnes capacity were installed for simultaneous operation of all the gates. The steel structure at RL 35 m for the sluices was completed between May 1991 and March 1992 (Fig. 4).

Image 3: Concreting for Raising a Pier

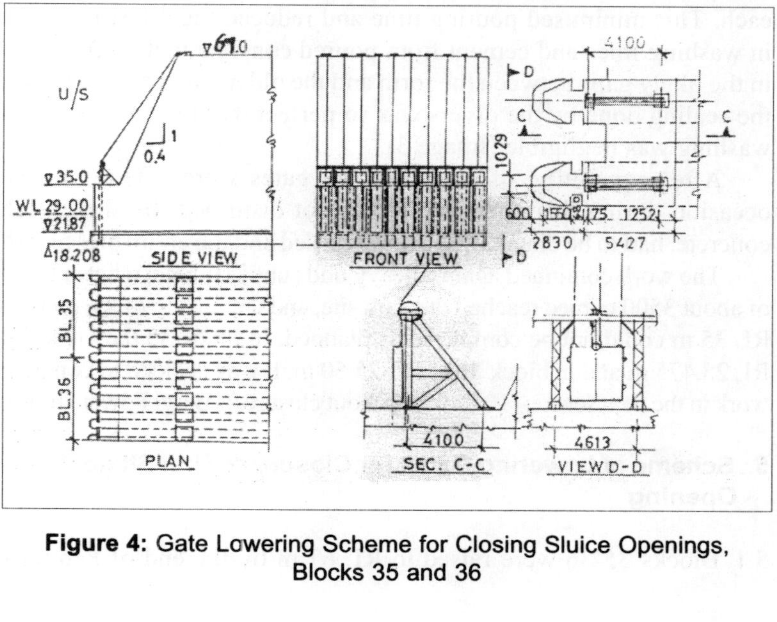

Figure 4: Gate Lowering Scheme for Closing Sluice Openings, Blocks 35 and 36

5.2 All the ten gates had a head gear arrangement at RL 35 m in front of the steel structure connected with winch rope. They were lowered on a trial basis to ensure their smooth travel in the grooves right down to the sluice sill levels. To confirm that this happened, the wire ropes used for lowering them were marked and checked during the operation. The grooves were cleaned with compressed air and a gang of four experienced divers were deployed to accomplish the task. Simultaneous lowering was necessary so as to avoid any differential head that might otherwise be caused.

5.3 The divers also caulked the gates on the sides and the bottom to prevent leakage. The reservoir capacity curve and the river flow suggested a time lag of 20-24 hours after closure of the gates for the water to rise from RL 23 m to RL 35 m since retrieval of all the gates might be difficult due to rise in water level, all the wire ropes were immediately cut and it was decided to keep all the gates closed under water so that plugging of the sluices could start immediately from the downstream end.

5.4 The reservoir capacity between RL 35 m and RL 60 m is 222 MCM. With the river flow during April at 300 cumec, it was assessed that water would start spilling through the sluices (i.e. the final stage diversion) in about eight days, and that the sluices would discharge 300 cumec at 7 m head, establishing reservoir level at RL 60 m. Four sluices in block nos. 44, 46, 48 and 50 at sill level 53.0 m were expected to keep inflow equal to outflow, maintaining a constant reservoir level and allowing work on bay no. 2 stilling basin in the next working season.

6. Plugging of Sluices

6.1 After closure of the gates, dewatering and cleaning of the construction sluices on the downstream side began. The first task was to plug with concrete one-third of the length of sluices on the upstream side and, if time allowed, to further plug the remaining length by June 1994. However, the task was completed for all the sluices before May 1995.

7. Status of Works

7.1 The dam profile at end of June 1994 is shown in Fig. 5. By this time, service bay nos. 1, 4, and 5 were also completed along with right training walls up to RL 40 m. The work in RTW above RL 40 m and that in stilling basin No. 3 continued. The excavation of the tail channel, the coffer dam and the proposed submersible bridge across the newly excavated tail channel were completed before April 1993 in order to allow the water to pass through the permanent sluices at RL 53 m and safely join the river without submerging the working area. The final stage diversion scheme as envisaged and accomplished is shown in Fig. 6.

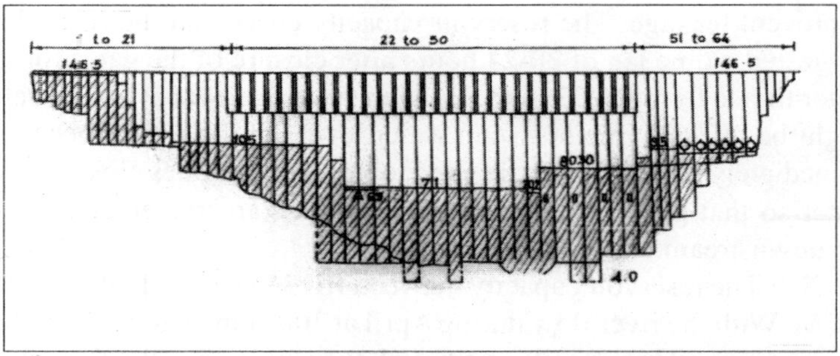

Figure 5: Profile of Dam at the End of June 1994

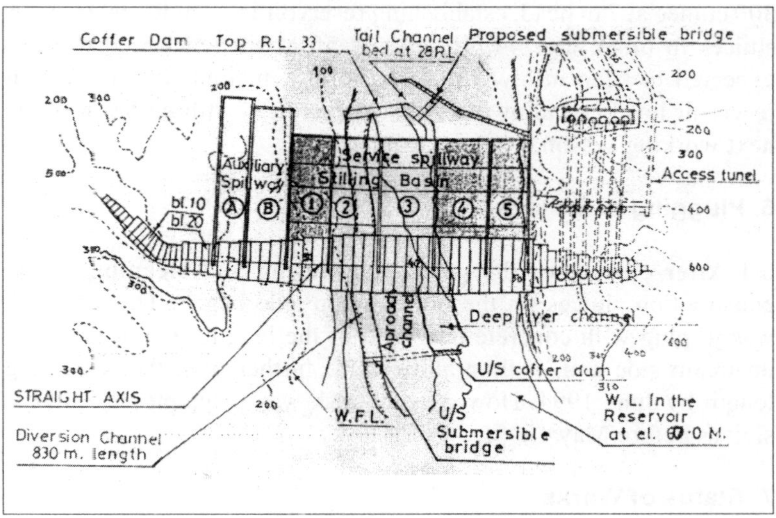

Figure 6: Final Stage Diversion Scheme

PART 2
Planning for SSP

7

Memoirs of a Narmada Planner

Mahesh Pathak[1]

Introduction

The Sardar Sarovar Project (SSP) is regarded as the most investigated/ studied irrigation project in India. It also had the distinction of being the most controversial project in India. The following account, based on this author's 15 years of close association with Narmada Planning Group (NPG), narrates how professional back-up helped in the evolution of this mega-project. It also discusses major policy issues relating to Municipal and Industrial use of water (M&I), Economic Appraisal, Resettlement and Rehabilitation (R&R) and Environmental Impact Assessment (EIA).

I. Narmada Planning: Early Investigations

Sardar Patel, soon after joining the interim government in 1946, had stressed that (the then two 'provinces' of) Bombay and Central Provinces and Berar should think of making the best possible use of the thitherto unutilised Narmada waters. At the request of the two provincial governments, the Central Waterways, Irrigation and Navigation Commission (CWINC) started studying the topography

1. The author wishes to thank V.D. Shah, Research Officer, AERC, Vallabh Vidyanagar for proof reading and Bhoomika Padhiyar, AERC, Vallabh Vidyanagar, for Computer Assistance.

and hydrology of the Narmada basin in 1947. After prolonged investigations and discussions, the Government of Bombay put forward a proposal for Navagam dam at FRL 320 ft.

Three months after the formation in May 1960 of a separate state of Gujarat, the Planning Commission accepted the proposal for Stage I of the Broach Irrigation Project. The project envisaged a dam at Navagam, to be built in Stage I to FRL 162 ft. with a low level canal taking off from there to irrigate 389,000 hectares. Stage II envisaged raising the dam height to FRL 320 ft. irrigating a further 364,000 hectares. After administrative approval by Government of Gujarat (GoG) in February 1960, Prime Minister Nehru inaugurated the project in April 1961. Detailed geological investigations followed for the selection of a more suitable dam site. In November 1963 the Chief Ministers of Gujarat and Madhya Pradesh (MP—the successor 'state' to CP and Berar Province) arrived at the Bhopal Agreement which provided for Navagam dam an FRL of 425 ft. Although GoG ratified this agreement and submitted a project proposal to GoI (30 November, 1963), the Government of MP objected to the Bhopal Agreement insisting on a dam height for Navagam of FRL 162 ft. only.

In order to resolve the dispute, GoI appointed a High Level Committee of eminent engineers headed by A.N. Khosla (5th September, 1964). The terms of reference for this expert group were decided in consultation with the state governments of Gujarat, MP and Maharashtra. The Committee was expected to draw up a plan for the optimum and integrated development of Narmada water resources. It submitted a unanimous report to GoI in September 1965, consisting of a Master Plan that envisaged 12 major projects in MP and one in Gujarat at Navagam. The optimum FRL for the terminal dam at Navagam was suggested at FRL 500 ft. In addition, the installed power generation capacity was to be 1400 MW. The Khosla Committee had emphasised that a higher Navagam dam would minimise wastage of water to the sea and permit greater carry-over capacity, increased power generation and assured optimum irrigation and flood control.

What was particularly noteworthy in the Khosla Committee report (Khosla, N.D.) (Ch. XI) was its approach to Narmada development planning. The basic guidelines provided for preparing the Master Plan emphasised that national interest should have overriding priority irrespective of state boundaries. Rights and interests of the states should be protected keeping in view the wider national perspective. Another noteworthy guideline was the top priority given to irrigation which should be extended to the maximum possible area and in particular to areas along the borders of Gujarat and Rajasthan, which would encourage sturdy peasants to settle in these border areas.

While Gujarat broadly endorsed the recommendations of the Khosla Committee, MP and Maharashtra raised strong objections, especially regarding the height of the Navagam dam and the control over distribution of the Narmada waters. When earnest efforts of the GoI (for about six years) for a mutually acceptable settlement failed, GoG submitted a petition to GoI (July 1968) for appointment of a Tribunal under the Inter State Water Disputes Act, 1956.

GoI accordingly constituted the Narmada Water Disputes Tribunal (NWDT) in October 1969. Gujarat's case was ably represented by a team of legal luminaries led by J.M. Thakore. It was pitched against equally competent teams of legal stalwarts, one led by F.S. Nariman for Maharashtra and the other by Y.S. Chitale for MP. The Gujarat team of irrigation engineers, led by C.C. Patel, included P.A. Raj, I.M. Shah, N. Ramaswami, M.M. Shah and N.B. Desai. It was this team that had put in a marathon effort of sustained hard work to present the Gujarat case before the Tribunal.

Even a cursory look at the proceedings of NWDT will reveal that the Tribunal was demanding, year after year, both a huge volume of data and technical studies relevant to Narmada basin development. Its deliberations continued over a period of 10 years, and it gave its report in August 1978 (Narmada Water Disputes Tribunal 1978). The final orders and decisions of the NWDT were gazetted on 12th December, 1979.

Knowledgeable irrigation engineers consider the NWDT award of great historical significance. It is regarded as a model award, which is so precise and well drafted that there is no scope for any ambiguity on sharing benefits and costs. This landmark award had three noteworthy features:

1. It was path-breaking as it adopted a national perspective for harnessing water resources. It recommended equitable and fair distribution of available water and power benefits.

Although Rajasthan was not a riparian state, the Tribunal had allocated 500,000 acre feet of water for drought prone Barmer and Jalore districts which had no dependable sources of water.

2. It recommended appropriate institutional mechanisms to ensure smooth implementation of and effective compliance with all its provisions. These included (a) the Narmada Control Authority (NCA) for carrying out decisions of the Tribunal, (b) the Review Committee to review any decision of the NCA, and (c) the Sardar Sarovar Construction Advisory Committee (SSCAC) for ensuring efficient execution of unit-I (dam and appurtenant works) and unit-III (power complex).
3. The tribunal was much ahead of its time in prescribing a liberal package of land for land and other living amenities for R&R of oustees.

The World Bank and NPG

GoI had sought World Bank (WB) assistance for financing the Sardar Sarovar Project (SSP). Following the NWDT award, the Bank had mounted a reconnaissance mission in November 1978 to review the Narmada basin development plans and to determine appropriate means for WB involvement. This mission had made three important recommendations (World Bank, 1985):

(a) GoG to establish a high level Narmada Planning Group.
(b) Retain private or independent consultants for systems studies and general planning of the GoG.
(c) Key foreign experts should be provided through the UNDP and special project preparation component of an IDA credit. World Bank, Staff Appraisal Report—Narmada River Development—Gujarat, February 12, 1985.

Following the WB recommendations, GoG constituted two important bodies in April 1981: an apex body Narmada High Power Committee (NHPC) under the chairmanship of the Chief Minister, constituted to provide overall guidance and direction for the project work; and the Narmada Planning Group (NPG) under the chairmanship of the Minister of Irrigation (the other members of NPG are listed at Appendix 1).

About two years later, S.R. Hashim left the NPG. I was offered, the post of Full-time Executive Member of NPG in November, 1983. I was willing to serve part-time, while continuing to work in Vallabh Vidyanagar. I met H.M. Patel (ICS Retd.) to seek his advice about joining NPG on a part-time basis. His cryptic remark was, "We must support this project work, which is so important for India". He further advised me that to work part time GoG must be assured that I would spare all the time required for attending to the NPG work. Upon my giving such an assurance an order of my posting was issued, and I joined NPG in May 1984 with an initial offer from the government of one year, which it continued for 15 years (1984-1998). I kept my promise till the end of my tenure, spending the prime time of my life for a mega-project so important for Gujarat and for the country.

On the technical side, NPG had some top class senior irrigation engineers including P.A. Raj, I.M. Shah, N. Ramaswami, M.M. Shah, N.B. Desai, M.U. Purohit, M.R. Goswami, B.J. Parmar, G.L. Java and others. The World Bank deputed 11 missions during Identification and Review phase (1980-81), four for Pre-Appraisal (1981-82) and five during the Appraisal Phase (1983). Some of the leading experts who helped the planning process included P.L. Jung, G. Fauss, C. Perry, M. Barber, L. Shanon, C. Diewald, and T. Scudder. After formal approval of support, W.B. Missions visited Gujarat every six months. The meetings with these missions kept the administration alert and they provided useful feedback both for policy and project implementation. Withdrawal of WB support later deprived us of the benefit of regular monitoring and evaluation by recognised experts.

Even before the NPG was constituted (April 1981) the Irrigation Department of GoG had initiated (1979-81) several studies on regionalisation of the command, main canal systems, interregional allocation of water, cropping pattern and water allocation, critical zones of the command and rehabilitation of Gujarat submergence villages.

Immediately upon the final award of NWDT in December 1979, GoG prepared the Sardar Sarovar (Narmada) Project report by January 1980 and an Identification Report on this basis, which was submitted to WB in July 1980. Simultaneously, WB, on the basis of its own

Identification Missions, produced and forwarded to GoI and GoG, in November 1980, a two-volume document titled "Planning for Narmada Development (Gujarat)." Volume I covered "The Planning Framework," spelling out the planning needs for such a mega-project. Volume II contained "Technical Annexes" providing detailed parameters of some of the major studies required for the project. Detailed discussions on the Planning Framework followed between the officials of WB, GoI and GoG, during 1981. Based on this spadework, NPG prepared a document on "Projectised Studies" for the SSP (Narmada Planning Group, 1981).

NPG, in consultation with the GoG Irrigation Department, undertook 17 technical studies covering diverse subjects such as reservoir simulation, dynamic programming, groundwater simulation, groundwater investigations, supplies from *en route* rivers crossed by the Narmada Main Canal (NMC), canal losses, CD works, operation plan for system-based transients, operation plan for the distribution system of sample area and special problems of NMC. The NPG also decided to undertake 29 socio-economic studies on a variety of subjects such as cost of cultivation, regionalisation of the Narmada command area, population projections, migration and settlement patterns, land-use and cropping patterns, water requirement, energy requirements, regional allocation of water, estimation of agricultural outputs and incomes, project impact on state economy, project costs, economic appraisal, rehabilitations and resettlement, ecology and environment, benchmark studies on socio-economic status and consumer expenditure and investment pattern in the command area.

The outcomes of many basic studies initiated in the first phase of work were incorporated in the project report "Sardar Sarovar (Narmada) Development Plan," which was presented to WB in May 1983. Volume I of this report (14 chapters) contained the outcomes of various studies and described the main features of Gujarat and its agriculture, SSP profile, water management planning, command area development and economic appraisal of the project. Volume II (also 14 chapters) mainly contained engineering and technical details of the project.

GoG wanted SSP to be planned, designed, constructed and operated at the highest standards of organisation and management. The planning framework, as it evolved, represented a break from the past in several important ways. Firstly, the project was expected to ensure equitable and timely supply of available water in the extensive command area demarcated. Secondly, through appropriate water management strategies, it was expected to avoid the problems of waterlogging and salinity usually associated with such large projects. Thirdly, it was expected to make fair and equitable arrangements for the R&R of project oustees. Finally, sustainable development was to be ensured by paying adequate attention to issues of environmental concern. WB noted in its Appraisal Report (February 12, 1985): "Within a period of three years an impressive array of high quality studies and designs have resulted, including a comprehensive framework plan by the NPG and production by the ID of high quality designs, specifications and procurement documents. GoG has prepared a project that represents a break with past approaches to the planning, design, construction and operation of irrigation projects in India" (World Bank, 1985). While the WB experts were lauding the merits of the SSP planning framework, there were many critics trying to prove it as a planning disaster. It was, therefore, decided to publish the main features of the SSP planning framework prepared by NPG. The title of this publication, 'Planning for Prosperity', reflected both the vision and ultimate goals of those who planned for this mega project (Narmada Planning Group, 1989).

Mode of Working of the NPG

Whether NPG was an autonomous body, as in the earlier stage, or a part of SSNL (after its formation in April 1988) it always enjoyed a high status and autonomy that a professional body requires for its working, as also did the experts undertaking studies for SSP. NPG enjoyed sufficient freedom in identifying the agencies/experts to whom studies were to be assigned. Project proposals were carefully worked out within NPG, each defining specific objectives, time frame and budget. NPG expected project teams to submit draft reports, which were thoroughly examined and comments/suggestions offered for finalising them.

Looking back on my long tenure with NPG, whichever be the party in power and whoever the Minister (I worked with Keshubhai Patel, Purushottam Rupala, Babubhai Jashbhai Patel and Jay Narayan

Vyas), at no stage was there undue ministerial interference in NPG work. Y.K. Alagh was formally in and out of NPG due to important assignments. However, he had an abiding interest in our work and always provided valuable inputs as and when required. The top level team of irrigation engineers remained largely the same, working with exceptional dedication. However, senior IAS officers were shunted around as per government directives; many of them were admirable for their professionalism and dedication. The top level irrigation engineers and bureaucrats always upheld the autonomy of the NPG. All of them were supportive and an atmosphere of camaraderie and togetherness prevailed. Although my job was daunting, this congenial atmosphere made it enjoyable and memorable.

The paragraphs that follow describe how NPG tried to provide an authentic, professional backup for this major project.

II. Municipal and Industrial Use (M&I)

Some anti-dam activists were questioning the intention of GoG to provide drinking water through SSP. Their contention was that although GoG got the Narmada project approved for serving the needs of water-starved regions, it would never quench the thirst of people living in Saurashtra and Kachchh. The promise of providing drinking water was termed as a myth. However, this misinformation campaign itself proved to be a myth, as drinking water supply turned out to be the first major benefit of the project. It was difficult to understand how and on what basis this claim was made, especially when the NWDT Award had specifically given 1.06 MAF of water for domestic and other non-agricultural uses on full development. Moreover, the Gujarat Water Supply and Sewerage Board (GWSSB) had prepared a detailed Master Plan for water supply from the Narmada (GWSSB, 1983).

The NPG publication 'Planning for Prosperity' had a separate chapter (no. 8) on Water Use Plan and Sizing of the System. The NPG planning framework had estimated total water requirement for domestic, industrial and thermal power projects at 1.37 MAF. It was also clearly stated that this total requirement would be met from available local sources (0.48 MAF), from recharge withdrawals (0.14 MAF) and from Narmada waters (0.75 MAF). When adjusted for losses in the system, the gross demand on the Narmada canal system worked out at around 1 MAF (Narmada Planning Group, 1989).

One of the most important accomplishments of SSP has been the planning and implementation of drinking water project which, when completed, is expected to provide benefits to 9633 villages and 131 towns in 17 districts of Gujarat. A major part of these towns and villages are in Saurashtra, Kachchh and North Gujarat, regions known for their acute water shortages. The Narmada Water Supply Project is regarded as the largest engineering intervention in India's rural water supply system. It is difficult to impute any economic or social value to this supply of potable water which will significantly improve the quality of life for the water-starved people. According to the latest available information from SSNNL, 114 towns and 6513 villages were already covered for this benefit by March 2010. It is noteworthy that measures to supplement water resources have also been taken. In the 15 districts of the command area, work on 159,957 *khet talawadis*, 70,657 check dams, 71,168 *bori bunds* and deepening of 12,672 village ponds had also been completed by 2009 (SSNNL, 2009).

What is important is that a definite work plan is in place and its implementation is well on course. Operational problems are bound to arise in a project of this size and complexity. It is hoped that a prompt responsive mechanism will help in resolving these issues as and when they arise.

GoG and the project authorities have attached a great deal of importance to the NWDT Award regarding water allocation for M&I. Considering the Narmada waters as a soft option for water supply, delegations of urban metropolitan authorities often visited the SSNNL with a request to make larger provision for their drinking water needs. Whichever the party in power and whoever the chairman of SSNNL, such demands were always politely but firmly turned down.

Almost three decades have passed since the NWDT Award was given and two and a half decades since the project planning framework was evolved. Inordinate delay, for whatever reasons, in the completion of the project was not anticipated. Over the years the earmarked command area for providing irrigation benefit has declined especially near the urban centres and near the industrial growth centres. The pace and pattern of industrial development in Gujarat during the past two decades was not anticipated when the SSP planning framework was evolved. The SSNNL Water Plan stated in bold letters and in unambiguous terms that "the industrial water usage from the project remains unchanged." (SSNNL, 2009: 6)

The GoG and the project authorities will decide the water use plan keeping in view the NWDT Award. However, it is necessary to mention here two policy related issues. First, the Project Completion Report of the World Bank has stated, "It is reasonable to expect that the water supply demand would continue to increase over time and the SSP could become a major water supply project. This is a worldwide trend. In this context, it would be prudent to consider reducing the command area and earmarking more water for the high-value D, M&I needs." (World Bank, 1993: 42)

The project authorities have been often criticised for catering to the needs of industry which may adversely affect the availability of water for irrigation. In support of this argument, the NWDT Award is also quoted which had made an allocation of 9 MAF of water for Gujarat after a careful review of its requirement of water for irrigation and M&I uses. However, it needs to be clarified that out of the 16 Clauses of Final Order and Decisions of the NWDT, Clause III was on apportionment of the utilisable quantum of Narmada Waters. Point (3) of this Clause states, "Within its share of water, each party state is free to make such changes in the pattern of water use and in the area to be benefited within or outside the Narmada basin in its territory as it may consider necessary." The Supplement Data Volume of Staff Appraisal Report of WB also mentions that "the NWDT Award left it to the party states to decide what use they wanted to make of their allocated shares." It further states, "Each state can use its shares for whatever purpose and in whatever part of its territory it deems fit." (World Bank, 1985: 94)

III. Economics of SSP

Economic appraisal of SSP has been carried out by several independent national and international experts. To begin with, NPG had taken the view that this appraisal had to be carried out by an agency of international repute so as to establish its credibility. Hence, it was decided to invite Tata Economic Consultancy Services (TECS) to prepare the project appraisal report. The TECS submitted to the NPG a 78-page document elaborating the methodological framework to be adopted for this project appraisal (TECS, 1981). This framework had been carefully reviewed by the concerned experts and further course of action was chalked out. Following the usually adopted standard methodology, TECS tried to cover, to the extent feasible, both direct and indirect costs and benefits of the project. It submitted its report to NPG in May 1983 (TECS, 1983).

Most of the data required for this study were generated internally or from studies undertaken by NPG. Though no part of the NPG team when this study was entrusted to TECS, I acquired first-hand knowledge about the experts' conceptual clarity and analytical skills when they approached Agro-Economic Research Centre, Vallabh Vidyanagar for data on cost of cultivation in Gujarat. Reliable data, required for deriving net irrigation benefits, were not available. TECS needed cost data generated through Comprehensive Scheme to study Cost of Cultivation in Gujarat, which they considered "the most detailed, extensively covered and scientifically collected data for the state." Project costs and benefits were estimated both at market prices and economic prices, broadly following the Little-Mirrlees Method. Based on a very comprehensive and systematic exercise, the TECS report said that "unequivocally, the proposed project is economically viable." The report indicated that SSP would provide much larger benefits than the costs involved with a benefit-cost ratio of 1.84 and IRR of 18.3 per cent both evaluated at economic prices. Ridley Nelson, the brilliant WB economist, later said in his memorandum of 1990, "I should emphasise that in my view, the original 1984 analysis is arguably the most sophisticated and most thorough ever undertaken by the Bank in agriculture (and I have probably seen more analysis than most staff have because of my reviewing projects in OPS)."

The World Bank Appraisal Report had estimated the economic rate of return (ERR) at 13 per cent (World Bank 1985). It had also carried out sensitivity analysis taking into account various favourable and adverse factors. This analysis yielded an ERR in the range of 7 per cent to 19 per cent; even the lowest figure was considered adequate. The Nitin Desai Expert Committee, constituted by the Planning Commission, had suggested a rate of return of 9 per cent

(7% for drought prone areas) for irrigation projects which is a real rate of return (Nitin Desai Expert Committee, 1983).

The World Bank updated its economic appraisal in 1990, which confirmed that the original ERR of 12 per cent was correct. It again updated its appraisal based on 1991-92 prices. The revised analysis, completed in April 1993 as part of the benchmarks set by the WB Board of Directors, is based on irrigation, power and M&I benefits including fuelwood and fisheries benefits, taking into account changes in quantities, prices and phasing since it was originally done in 1984. This analysis paid particular attention to environmental costs and benefits (World Bank, 1993). The revised ERR was estimated at 12.2 per cent, which was marginally lower than the original 13 per cent. The report considered this within acceptable range for a project of this type "exhibiting a very large scale of benefits relative to any feasible alternatives and having substantial multiplier effects."

The WB Project Completion Report for SSP did recognise that the "robustness of ERR calculations will be clouded by many uncertainties". However, despite the uncertainties of timing and efficiencies, the World Bank did not consider the base case ERR estimate to be skewed in the distribution of possible outcomes. On the contrary, it indicated that the real benefits might be higher than projected as the benefits of power and M&I had been undervalued. It also suggested that as the cropping pattern of SAR was conservative, introduction of more high value crops and gradually increasing water use efficiencies through adoption of micro-irrigation may provide higher agriculture benefits.

Manu Shroff, a distinguished economist well versed in public policy, was invited to contribute a review paper on economic appraisals of SSP to a seminar at Vallabh Vidyanagar. He argued in his paper (Shroff, 1995) that WB has been in the business of lending and its economic appraisals are bound to be cautious and conservative. It would hardly agree to finance a project with "dubious economics." The WB experts, who had acquired intimate knowledge about the project over a decade, had all along (right from the first appraisal report to the last project completion report) upheld the economic viability of SSP.

One of the benchmarks laid down by the Bank was to review with the project authority the methodology and parameters for updating the economic appraisal. Keeping in view this development, NPG invited the Ahmedabad based Sardar Patel Institute of Economic and Social Research (SPIESR) to update the economic appraisal earlier done by TECS. This exercise was also done at 1991-92 prices. Again the IRR worked out with three different assumptions of water use efficiency (61%, 53% and 45%) and project costs (Rs. 153, Rs. 132 and Rs. 110 billion) was estimated in the range of 16.77 per cent to 21.88 per cent (Alagh *et al.,* 1995). This study expected the dam to be completed by the year 2000, canal work by 2002 and the River Bed and Canal Head Power houses by 1999. The M&I water benefits were expected to start from 1995-96 and reach peak level by 2005-06. This economic appraisal has been partly validated by the flow of power and M&I benefits. Its economic viability will be further confirmed when the visualised irrigation benefits from this project become available.

Project Financing

As part of the World Bank Agreement, GoG was required to undertake a study on water rates policy for SSP which NPG arranged to conduct (Rao, 1992; H.G. Patel, 1990; Patel and Patel, 1992). Part I of this study presented a comprehensive review of literature on the pricing of public utilities with special reference to irrigation water. Part II provided useful insights by examining the financial working of four irrigation projects (Dantiwada, Kakrapar, Mahi, Shetrunji) in Gujarat. Part III examined important policy issues relating to canal water charges, which was intended to help GoG in formulating an appropriate water rates policy.

Chaman Kumar, IAS, Executive Director (Finance) SSNNL had prepared an internal note during 1995 that had expressed concerns about several aspects of project financing. In this context, I requested Manu Shroff, who was well versed in the public finances of GoI and GoG, to help in examining this subject. He agreed and Jay Narayan Vyas, then Minister, Narmada promptly accepted a proposal to constitute an expert group to review the financing of SSP. NPG constituted an Expert Committee with Manu Shroff as chairman, Chaman Kumar as member (subsequently Rajiv Gupta) and me as convener. The expert group was constituted in December 1995. It submitted its report in August 1996 (Narmada Planning Group 1996). We had put in a lot of effort over a period of six months to prepare this report, to which Manu Shroff had made valuable contributions. He politely declined to accept any remuneration for his services.

IV. Resettlement and Rehabilitation (R&R)

According to the latest information made available by the Sardar Sarovar Punarvas Agency (SSPA) regarding the status of R&R as of 31 July, 2010, out of 46,732 project affected families (PAFs) belonging to 244 affected villages, 46,330 have been resettled (11,107 in Gujarat, 772 in Maharashtra and 32,192 in MP). Gujarat, apart from settling its own oustees (4,753) has also provided R&R benefits within Gujarat to oustees from Maharashtra (772) and MP (5,582).

This achievement is bound to come as a pleasant surprise to those associated with the earlier phases of turmoil and struggle. Even some of the WB experts, who were all along supportive of SSP, thought at some stage that while R&R in Gujarat would be satisfactory, the challenge of MP oustees would be so daunting as to be almost intractable. NWDT had taken a sympathetic view of the oustees and the norms it recommended for R&R were perhaps the most liberal ever laid down for an irrigation project in India. It had demanded details of past R&R efforts in major irrigation projects. After examining this information it issued specific and elaborate directives for R&R in Clause XI of the final order. ARCH-Vahini was pushing for some refinements in the R&R package to fill up some important gaps. It was around this time that WB deputed Thayer Scudder, an international expert, to review the status of R&R in SSP (November 1983). Later, as a member of the WB team that negotiated the loan agreement, Scudder helped further refine the R&R package (May 1985). It was then that GoG proclaimed its R&R package as one of the most liberal ever to be provided in India. In brief, the R&R programme was expected to ensure for oustees, within a reasonable transition period, living standards and conditions equal to or better than those they were in before displacement. Social and economic integration of the oustees with the host community was also to be ensured as far as possible.

There were aggressive groups of activists in India and elsewhere who were totally against big dams and who questioned the intention and ability of the government to ensure the promised benefits. In ARCH-Vahini, led by Anil Patel, Gujarat had an equally aggressive and yet liberal and pragmatic group which spearheaded a relentless battle for a fair and equitable R&R package. Once such a package was declared, this group focused attention on its proper implementation. The drama of R&R for SSP as it unfolded had many actors on the stage—PAPs, GoI and concerned states, WB, the national and international NGOs and the professional experts. For those interested in this drama, ARCH-Vahini (1995) and Sudarshan Iyengar (1997) have given a detailed account. Some people may find their presentations blunt and aggressive, but their commitment to the cause of social justice is irreproachable.

In the early 1980s, when work on SSP was initiated, it had been widely recognised that previous efforts of irrigation development, both in developed and developing countries, had not been adequately guided by norms of social justice and equitable distribution of the benefits of development. The NWDT Award was historic in that it issued specific directives for R&R benefits of land, housing and other amenities to PAPs. It was binding on the party states. The SSP planning framework had not incorporated a detailed blueprint for the R&R programme implementation. Similarly, none of the participating states had anticipated the enormity and complexity of the problem. Hence the responsive and proactive administrative machinery required to handle it was not in place. The need to handle this human problem with understanding and empathy was not fully realised. Thus, the initial efforts for R&R had met with stiff resistance from several NGO including ARCH-Vahini.

Gujarat had arranged to conduct a socio-economic benchmark survey of 19 villages facing submergence, which was used by CSS, Surat for a comparative study of pre- and post-project socio-economic status of oustees. The NGO agitation for an equitable R&R package and its effective implementation was mounting in an unprecedented mass mobilisation of oustees to assert their demands. Sanat Mehta, then Chairman of SSNNL, skilfully negotiated the controversial issues with ARCH-Vahini. Once the R&R package was accepted by GoG, ARCH-Vahini realised that things were moving in the right direction, even if slowly. It advocated continuous and effective oversight for implementing the R&R programme. The credit for such a model R&R policy goes to the NWDT Award, a powerful NGO movement led by grassroots social activists, WB guidelines, independent monitoring and evaluation by experts and valuable support from some enlightened politicians and bureaucrats.

By the end of the 1990s both WB and GoG recognised that the enormous task of R&R could not be handled through regular administrative machinery. This led to the establishment of the Sardar Sarovar Punarvas Agency (SSPA) in 1992 for speedy and effective implementation of R&R package.

The initial benchmark survey of 19 submerging villages of Gujarat was prepared by Vidyut Joshi. NPG assigned the task of monitoring and evaluation (M&E) of SSP's R&R programme to the Centre for Social Studies (CSS), Surat, which was engaged in this work during the entire length of my association with NPG (1984-98). CSS submitted 24 six-monthly draft M&E reports, based on their field experience and in consultation with NPG and the R&R Cell. SSNNL reviewed and commented on them before their finalisation. The regular M&E reports were prepared by Ghanshyam Shah, S.P. Punalekar, Nancy Lobo and D.C. Sah. In the initial years, when R&R work was being organised, many implementation problems were reported. Meetings to review draft reports were stormy, with officials defending the government work and CSS experts emphasising their independent assessment. An interesting fallout of this M&E was that as soon as the draft reports came in some of the issues raised were promptly attended to before the review meetings to discuss the reports.

As GoG geared up the R&R work, the M&E reports for the later period (Nos. 17-24) provided a brighter picture of the benefits of various R&R programmes. During this period I read an important paper by Michael Cernea, a WB expert on R&R (Cernea, 1995). He had identified from his extensive research in R&R several dimensions of impoverishment risks for displaced people. He found that impoverishment due to displacement was primarily in terms of landlessness, joblessness, homelessness, marginalisation, increased morbidity and mortality, food insecurity, loss of access to common property and social disintegration. But he did not regard such impoverishment as being inevitable. Given equitable policies and programmes, the impoverishment risk model could be transformed into a self destroying prophecy. Based on the data available in CSS R&R reports (Nos. 17 to 21) I had prepared a note to test the Cernea thesis.

Report Nos. 17, 18 and 21 provided data on the status of basic amenities such as education, health and transport. These were significantly better at the resettlement sites as compared to the submerging villages. Domestic and street lighting, totally absent in the submerging village, was available at the new sites. Crude death rates and infant mortality rates were reported to be significantly lower in the new sites (Report 18). Average land owned by the oustees was 5.7 acres as against 2.75 acres in the submerging villages (Reports 19-20). Reports 19-21 found a relatively more diversified cropping pattern at the new sites. Significant improvements in crop yield (kg per acre) were also reported: from 95 to 220 for maize, from 95 to 190 for jowar and from 87 to 154 for cotton. Yield of paddy and oilseeds, not cultivated in the submerging villages, was 379 kg and 182 kg respectively in 1993-94. The reports also found increased expenditure on consumer durables (four-fold), education and health (seven-fold), food (three-fold) and household assets (seven-fold).

Around this time some people criticised the GoG's R&R measures favouring Gujarat oustees and neglecting the MP oustees. NPG, therefore, requested Sudarshan Iyengar to take up a study of MP oustees resettled in Gujarat (Iyengar and Dave, 1998). Iyengar, a leading social activist, and his team visited 40 sites in Gujarat where MP oustees were resettled. This report also indicated that most of the sites had all the basic amenities, though some problems about drinking water and housing were also reported. The health services at the new sites were generally found much better. More than 90 per cent of the land was cultivated by the settlers at the new sites and only around 7 per cent of the plots had problems of size, quality, location etc. At most of the sites, relations between the original inhabitants and new arrivals were cordial. A large number of settlers categorically reported that they were better off in Gujarat than in M.P.

This demonstrates the extent to which the impoverishment risks associated with such displacement have been eliminated. Recognising that satisfactory R&R was essential for the progress of SSP, GoG had given it priority. Several R&R review meetings were held by Chimanbhai Patel, then Chief Minister of Gujarat. The concerned officials would submit their progress reports and discuss the problems they faced. Leading NGO representatives and social activists such as Anil Patel, Chunibhai Vaidya, Krishnaprasad Patel, Sudarshan Iyengar, Vidyut Joshi and others would put forward field level problems demanding immediate attention. On-the-spot decisions were taken and directions given for implementing R&R on a priority basis.

However, it would be naive to presume on this basis that everything was fine with the R&R programme. A variety of problems, big and small, would crop up. GoG was making efforts to evolve a responsive mechanism to resolve these problems as and when they arose. As a Narmada planner I cherished the dream that SSP oustees, instead of becoming the victims of development, should be the first beneficiaries of this mega-project. Needless to say, this called for effective implementation of R&R policy with necessary understanding and empathy.

Looking at the big picture, fairly elaborate arrangements had been made for R&R. NWDT had issued specific directions which were binding on the states. The Ministry of Environment and Forests (MoEF), GoI was monitoring progress in terms of given parameters. NCA had a Review Committee and a broad-based R&R Sub-Group (with several independent experts) to monitor and ensure effective implementation. GoG had established a separate agency (SSPA) to ensure speedy and effective implementation of R&R. As per Supreme Court directive, GoG had established a Grievance Redressal Authority (GRA) headed by a retired Supreme Court judge (such authorities were also established in MP and Maharashtra). Any project affected person could approach this authority for redressal of grievances and its decisions were binding on the state government. It helped a great deal in resolving genuine problems and in bringing transparency and credibility to the R&R work. Apart from this, an independent M&E agency was reviewing the work and ever-vigilant NGOs were watchdogs for the programme.

V. Development Imperatives and Environmental Concerns

Soon after the NWDT Award in 1979, the GoG took up the planning of SSP in right earnest in consultation with GoI and WB. NPG had recognised the importance of environmental issues right from the beginning and a study on Ecology and Environment of the SSP was entrusted in 1983 to the M.S. University of Baroda covering the upstream and downstream environment of the project. This initial benchmark survey provided valuable data for further planning (Department of Botany, 1983).

However, in the early 1980s several environmental issues including R&R were raised. Critics were trying to brand SSP an environmental disaster. While some of the concerns were genuine, certain region-specific socio-economic imperatives were also important. Safeguarding Environment and Removal of Poverty had to be accepted as complementary objectives. The policy makers realised that the benefits of SSP large enough to outweigh the costs of immediate social and environmental disruption.

When the SSP planning framework was prepared, neither GoI nor WB had elaborate guidelines for integrating environmental issues with irrigation project planning. WB had recruited its first environmental adviser as far back as in 1969, but a major reorientation of its policy on environment came only in 1987 when it created a Central Environment Division and four Regional Environmental divisions. Hence more detailed Environmental Impact Assessment Studies (EIAS) were undertaken by NPG on the basis of comprehensive environmental guidelines evolved in the 1980s by the Bank, MoEF, the Central Water Commission (CWC), GoI and the International Congress on Irrigation and Drainage (ICID). MoEF gave environmental clearance as late as June 1987. The Planning Commission gave its clearance in October 1988.

The decade of 1980s was marked by ceaseless attacks on SSP on environmental issues. It was during this period of turmoil that I delivered the presidential address at the 19th annual conference of the Gujarat Economic Association held at Bhuj in February 1989. I selected Environmental Planning for SSP as my theme. In this address, while recognising the importance of hitherto neglected environmental issues, I tried to establish the significance of SSP for Gujarat and how its various unique features took care of environmental hazards. I pleaded that all possible measures should be taken to minimise adverse impact and to maximise positive impact through abundant opportunities available for improving the environment.

While the crusade against the anti-dam movement was prolonged and nerve wracking, there were sporadic interludes of relief and amusement as well. I had sent a copy of my presidential address to Babubhai Jashbhai Patel. He suggested to the Chairman of SSNNL that it should be published for wider circulation (Pathak, 1989).

SSP was the first major irrigation project in India subjected to a rigorous environmental test, to prepare the required data base and to conduct all necessary EIA studies for integrating environmental concerns in its planning. The environmental planning framework focused on two major objectives. First, a modern, sophisticated irrigation management system was to be introduced that was both efficient and equitable. Second, it was expected to mitigate, to the extent possible, the adverse environmental impact of the project and exploit fully the ample opportunities that it offered for enrichment of the environment.

When the SSP planning framework was evolving, it was widely recognised that the traditional irrigation system suffered from two major limitations. First, it was not achieving the desired levels of water use efficiency, leading to problems of waterlogging and salinity. Second, the wider issues of environmental concern (including R&R) were neither fully recognised nor incorporated in the project planning framework. As will be evident from the following discussion, the SSP planning framework had paid adequate attention to both these vital issues.

Sustainable Water Management

NPG had initiated a number of studies to evolve an efficient and equitable water management planning framework. The range and depth of the studies is quite impressive. A comprehensive benchmark socio-economic survey of 62 blocks/talukas of the SSP command area was undertaken with the help of 14 different agencies including universities and research organisations of Gujarat (1982-83). As the command area was characterised by wide diversity in agro-climatic and socio-economic conditions, a regionalisation study was done by ORG, Vadodara (1982). Taking into account the annual rainfall, land irrigability class, water quality and quantity and drainage conditions, the entire command area was divided into 13 agro-climatic zones. ORG also did studies (1983) on cropping pattern and water demand.

The necessary soil surveys (rapid reconnaissance, reconnaissance and detailed) were conducted through National Bureau of Soil Survey and Land Use Planning. A long term study on groundwater investigation and monitoring was initiated through GWRDC. These investigations provided a dependable framework not only for ground water use planning and setting up of ground water models, but also for providing a sound scientific database for systematic monitoring of ground water levels and quality well before the project started water delivery in the command area.

Studies on the mathematical modelling of ground water were done through ORG, Vadodara (1982-83), Dalal Consultants, Ahmedabad (1993) and Consulting Engineering Services (CES), New Delhi (1993). These provided useful insights into the broad quantities and quality of ground water that could be integrated with surface water for conjunctive use. They also located areas vulnerable to waterlogging.

Efficient and equitable use of land and water resources has to be the major focus in irrigation project planning for sustainable development. In view of this, NPG mounted extensive studies and had detailed discussions with national and international experts to evolve an appropriate water management planning framework. This framework sought to provide irrigation benefit to as large an area as possible. Even though some WB experts were not in favour of providing irrigation benefit to special areas (Bara tract in agro-climatic region 4 and Bhal area in region 7), these areas were included in the command to ensure social justice. Studies were done for these regions and special irrigation strategies with limited water application worked out for such areas.

Anti-waterlogging and anti-salinity measures were inbuilt in planning, design and operation of the system. Some of the prominent features of the water management plan were: (1) limited water delta of 53 cm against the normal 75 cm in existing irrigation projects; (2) volumetric and rotational supply by *varabandhi*, (3) conjunctive use of surface and ground water; (4) provision of surface and subsurface drainage; (5) lining of main canal, branch and entire canal network up to 8 ha block to minimise seepage loss; (6) remote controlled automatic canal operation; (7) water balance and salt balance studies and monitoring; (8) encouragement of micro-irrigation; and (9) participatory irrigation management.

This water management plan was foolproof, leaving hardly any scope for criticism. While all other major features of the SSP were targeted by critics, the water management framework was the least criticised.

NPG invited a reputed Hyderabad-based consulting firm for a mid-course review of command area development planning. The firm submitted two interim reports in May and October 1993 and a final report in May 1994 (Wamana Consultants, 1994). These provided useful recommendations relating to project organisation, role of CAD and line departments, drainage and groundwater, training, monitoring and legal policy issues.

NPG took a noteworthy initiative to encourage the State Agriculture Universities to undertake a long tem research programme for evolving sound soil and water management technologies for sustainable development. These research efforts began well before the delivery of Narmada Waters. On the recommendation of NPG, SSNNL had agreed in 1988 to establish a new 22-hectare research station-*cum*-demonstration farm in black cotton soil at Khandha in Karjan taluka of Vadodara district. This research station was expected to conduct its research keeping in view the limited water for irrigation to be made available in the Narmada system. Following this initial effort, four other existing research stations at, Tancha, Dabhoi, Thasra, and Dhandhuka were also supported to undertake similar research. These studies were expected to provide valuable feedback for ensuring efficient use of land and water resources.

The early phase of project work faced no-holds-barred-attacks on almost all major issues related to the SSP. Today it is striking that SSP has succeeded reasonably well in all the major issues on which it was challenged and vehemently criticised. Work on water management and command area development has been delayed for reasons not anticipated earlier. Looking back at the water management planning framework, it appears that that a planner is in a sense a dreamer. He tends to incorporate, as he should, all desirable features in order to evolve a model planning framework. The stark reality is that what is desirable may not always be easily achievable. However, this cannot be an excuse for undue delays. Although the NPG's water management planning framework is fundamentally sound and the project authority must continue to strive to achieve the ideal, midcourse corrections and alternative back-up arrangements need to be made to keep the system working.

Environmental Assessment Studies and Compliance

During the early stage of project planning, a multidisciplinary group of experts from MSU had conducted a pioneering study on the ecology and environment of SSP (July 1983). It covered both upstream and downstream environs, collected systematic data on various ecosystem parameters and made a scientific and objective assessment of positive and negative impacts on the environment.

The Planning Commission also took an early initiative to examine the environmental aspects of the SSP and Narmada Sagar Project (NSP). As decided in a meeting held in May 1984, by Mohmmed Fazal, Member, Planning Commission, the Ministry of Agriculture, GoI, constituted an interdepartmental Committee in September 1984 under the chairmanship of H.L. Diwan. This committee studied the status of catchment areas of SSP and NSP and suggested detailed soil conservation and afforestation measures for the protection of the catchment area (Ministry of Agriculture and Cooperation, 1985). An integrated approach for catchment area treatment was suggested for implementation on a priority basis over period of 10 years. WB in its Staff Appraisal Report (Report Nos. 5107 and 5108, February 1985) had recognised the importance of environmental issues and suggested the need for studies and work plans relating to fish and fisheries, forest and wildlife, public health and irrigation and drainage network (command area).

Detailed guidelines for EIA studies were prepared by the World Bank and other international agencies as well as by GoI. Keeping in view the revised and more detailed guidelines, NPG mounted all the additional studies required for EIA.

Following the initial benchmark study in 1983, a multidisciplinary expert group of the Botany Department of MSU prepared a more detailed study on eco-environmental and wildlife management for SSP submergence area in Gujarat (Department of Botany, 1992). Detailed flora and fauna studies for the command area were prepared by the Bio-Sciences departments of the Sardar Patel University (area between the rivers Narmada and Sabarmati, 1995), Gujarat University (area between river Sabarmati and Rajasthan border, 1998) and Saurashtra University (area of Saurashtra and Kachchh, 1996). These studies suggested that no major adverse impact on flora and fauna was expected. Though the irrigation regime might in the long run bring about floral changes, they were not likely to be so drastic as to threaten particular floral and faunal species. Specific conservation plans may be necessary for some of the endangered species listed in the reports.

Eco-environment and wildlife management aspects of the Shoolpaneshwar sloth bear sanctuary in the downstream area were studied by the MSU team. NPG had also constituted an expert group to study environmental issues relating to this sanctuary. Follow-up work plans were prepared for implementation.

The MSU study (1983) had examined environmental impact on fisheries in the upstream and downstream area of SSP. A more detailed NCA sponsored study on aquatic life was undertaken by Central Inland Capture Fisheries Research Institute, Barrackpore (CICFRI N.D.). This expert review examined the environmental impact and also suggested concrete action plans for pre-impoundment and post-impoundment stages.

NPG constituted three expert groups to study the environmental issues pertaining to three sanctuaries in the command area—the Little Rann of Kachchh Wild Ass Sanctuary, Nalsarovar Bird Sanctuary in Region 7A near Sanand and Blackbuck National Park at Velavadar in Region 7. In order to facilitate the work of these expert groups, the Gandhinagar-based Gujarat Ecological Education and Research Foundation (GEER) was asked to prepare environmental status reports on the three sanctuaries. The expert groups visited all the sanctuaries to obtain first-hand information about the issues involved. Their recommendations were to conserve biotic, wildlife and endemic resources of the areas and provide Narmada waters to preserve and enhance their environmental value.

A multidisciplinary team of experts from the MSU Zoology Department undertook an NPG-sponsored EIA study on fisheries in the command area of SSP (Zoology Department, 1994). This systematic and comprehensive study evaluated environmental impact on existing and proposed potentials for fisheries development. It surveyed water bodies throughout the command area during pre-monsoon; monsoon and post-monsoon periods to prepare a data bank on soil and water qualities and plankton diversity. It suggested measures for exploiting the available potential and for mitigating the adverse impact.

Public Health

The importance of public health issues in irrigation projects have been recognised long back in the guidelines issued by the Ministry of Agriculture, GoI (1980) and subsequent guidelines prepared by MoEF, GoI (1985) and Central Board of Irrigation and Power (1986). The public health impact of SSP was first examined by the MSU in 1983. This study indicated schistosomiasis as a possible health hazard on the basis of experience of some African countries (Egypt, Sudan, and Zaire). Although it was not found in major irrigation tracts of India, as a matter of precaution, an intensive study was undertaken by a team of National Institute of Communicable Diseases, WB and WHO experts. This study concluded that schistosomiasis is not likely to occur in the Narmada Project region. NPG and GoG formulated an action plan on priority basis to look after the health related issues in the area.

NPG had also entrusted to the Directorate of Public Health, GoG, a study on public health in the command and downstream areas of SSP (Commissioner of Health, Medical Services and Education, 1993). This comprehensive study provided a detailed profile of health infrastructure and time series data on the incidence of various diseases. Based on the identified impacts, a health work plan was prepared for phased implementation. Although availability of water should help improve the health profile of people, hazards in terms of certain water borne/related diseases cannot be ruled out. In view of this, NPG emphasised the need for proper surveillance and control measures for malaria and other water related diseases.

Archaeological Sites

Large irrigation projects often involve submergence or relocation of archaeological, historical and cultural sites. As the Narmada is considered sacred in the Indian tradition, the governments of Gujarat, MP and Maharashtra had identified such sites in consultation with the Archaeological Survey of India and the concerned state departments. The NPG took due care of the environmental aspects of archaeological and historical sites in Gujarat.

In Gujarat, two temples going under submergence—Shoolpaneshwar and Hanfeshwar—have been duly relocated at new sites under the supervision of the Department of Archaeology. In the case of the command area the canals, big or small, can always be diverted or suitably realigned to ensure protection of important sites. NPG had obtained a comprehensive list of important sites both from the Archaeological Survey of India and Department of Archaeology, GoG. These lists were circulated to the concerned project officers to ensure that the sites were meticulously avoided in the planning and implementation of canal network. A large number of important religious places, temples, mosques and shrines—visited by thousands of people every year, are within or in the vicinity of the command area. The supply of fresh Narmada waters was proposed to revitalise these places.

Phase of Turmoil and Uncertainty

Even though conditional environmental clearance for SSP was given by GoI and GoG had agreed to fulfil the conditions, the anti-dam lobby continued its tirade against the project leading to considerable turmoil and uncertainty. The Supreme Court had put a ban on dam construction in 1992. How uncertain the atmosphere was can be judged from this example. In the early 1990s, one could easily anticipate that the anti-dam lobby would convince the people of Rajasthan that the Narmada Waters would not reach them. The situation was so uncertain and depressing that even some Rajasthan Government officials were sceptical about the supply of Narmada waters. Taking into account our experience in organising EIA studies, The Government of Rajasthan invited me to help them finalise the design of an EIA study for the Narmada command in Rajasthan. In a note prepared by Rajasthan Irrigation Department on potential sources of irrigation, Narmada waters were not accounted for. When asked for the reason, the officer said they were not sure when this project would be completed. Even if the dam was constructed, they were not sure if Narmada waters would ever flow into Rajasthan.

I vividly remembered this incident when I read the newspaper reports that Narmada waters had traversed 458 km in Gujarat to enter Rajasthan near Silu village of Sanchor Tehsil, Jalore district, on 27th March, 2008. The 74 km long Rajasthan stretch of the canal will pass through some of the worst drought affected areas of Barmer and Jalore districts, where people used to walk 10-15 km just to fetch drinking water. Nearly 100,000 people had gathered to watch Narmada waters entering Rajasthan and many women carried this water home in earthen pots to worship it. They understood the value of this water. For a Narmada planner, who was born and brought up in a water starved region of Gujarat, this was a moment of great joy and satisfaction.

It was encouraging to learn how Rajasthan planned to use this water. Earlier the state government had planned to irrigate 151,000 hectares of land from its NWDT allocated share of 0.5 MAF of Narmada Waters. However it has now modified the original plan with the help of Tahal Consulting Engineers, New Delhi, according to which the Narmada canal in Rajasthan will irrigate as much as 246,000 hectares (126,000 ha. through flow and 120,000 ha. through lift). Besides providing irrigation, the canal network will also provide drinking water facility in 1336 villages and three towns situated around the irrigation canal. The government and the people of Rajasthan have truly recognised the value of precious Narmada water. It would be worthwhile for Gujarat to watch their novel experiment of pressure irrigation through sprinklers. If it succeeds, Gujarat could think of emulating it wherever feasible.

The World Bank, embarrassed by the persistent and ever mounting pressure of the extensive international network of the anti-dam lobby, appointed an Independent Review Mission on SSP chaired by Bradford Morse in September 1991. The top management of the Bank was perhaps searching an escape route to wriggle out of this imbroglio. This Expert Group submitted its report to the World Bank in June 1992 (Independent Review Mission, 1991). Soon after the formation of this expert group, considering the attention that SSP was receiving both within and outside India, the Department of Economics and the Agro-Economic Research Centre of the Sardar Patel University organised an important seminar. D.T. Lakdawala, a distinguished economist who was then Deputy Chairman of the Planning

Commission, inaugurated this seminar, In his keynote address he analysed the importance of the project in the context of region-specific agro-climatic and socio-economic conditions prevailing in the project area. Based on his rich and varied experience of development planning in India, he had suggested that "all regional analyses of poverty problems have shown that when agriculture is growing and prosperous, the percentage of poverty is the least, and in a developing country no single factor contributes more to environmental deterioration than rapid population growth and persistent poverty." This seminar was attended by a large number of eminent professionals (economists, engineers, sociologists, political scientists, bio-scientists), senior administrators, policy makers and noted social activists. Reputed professionals presented papers on important aspects of SSP.

Soon after the publication of the Morse Committee report in June 1992, another important seminar was organised at Vallabh Vidyanagar in August 1992. The response to this seminar was overwhelming, with participation by more than 100 experts, policy makers, senior administrators and social activists. It was inaugurated by the Chief Minister of Gujarat, the late Chimanbhai Patel. In his inaugural address he declared in unequivocal terms that no power on earth would be allowed to stop the project. He also promised that all conditions relating to environment and R&R would be fulfilled. Since both these seminars provided documentation of historical significance, it was decided to publish their proceedings for wider circulation (Pathak, 1995).

Immediately after the release of the Morse report, the SSNNL published an appropriate response to it. GoG also published a fairly detailed response to this report (Government of Gujarat, 1992). Y.K. Alagh contributed a comprehensive and thought-provoking foreword for a publication containing proceedings of the two historic seminars held at Vallabh Vidyanagar during 1991-92 (Pathak, 1995). The Morse report did contain some useful insights and suggestions. However, it was generally regarded as a highly biased document which found almost everything wrong with the SSP. What was expected to be an independent mission had turned out to be a fault finding mission.

During the early nineties, critics were branding SSP as an environmental disaster claiming that it would lead to waterlogging, soil erosion, destruction of forests, wildlife and fauna and trigger health hazards for which environmental assessment and planning had not been done. NPG therefore thought it necessary to counter these proclamations with a factual presentation by publishing an authentic account of the EIA studies conducted and comprehensive framework evolved for operationalising and monitoring the management plan. This 356-page publication provided a synthesis of all environment related issues of SSP (Alagh, Pathak and Buch, 1995). The uproar on SSP environmental issues had largely subsided after this *bona fide* publication.

After the publication of the Morse report NPG had agreed to entrust some environment related studies to H.R. Wallingford, a UK-based consulting firm of international repute. WB had agreed to provide financial support for this work. Studies on hydrology, sediment and backwater, downstream area, agro-chemical pollution and command area EIA were entrusted to Wallingford in January 1993.

Wallingford in its report on the hydrology of SSP, noted that the river Narmada had been gauged at Garudeshwar, a few km below the SSP dam, since 1948 and at 32 other gauge stations upstream set up in 1948. After examining available records, it concluded that "the primary flow record is sound overall and fit for the purpose to which it has been put" (Wallingford, 1993). Wallingford had also analysed the dependable flows data and it had concluded that "Among the different segments of the series analysed by the CWC, the 100-year record is considered the most useful as a basis for allocating water shares among states. It yields a 75 per cent dependable annual flow of 32.9 BCM (26.6 MAF) which confirms the value of 27 MAF used by the Tribunal."

Wallingford in its report on sediment and backwater aspects of SSP had reviewed studies carried out in the past on the life of SSP and NSP and increase in flood levels upstream of SSP (Wallingford, 1993). It had concluded that "though some aspects of the analysis can be improved using concepts and methods evolved more recently, the studies provide reasonable estimates of the likely impacts." It also indicated that sedimentation would not be a major problem if NSP were also built.

Wallingford had examined the downstream impact of SSP within the framework of the ICID Environmental Checklist of possible impacts (Wallingford, Environmental Changes Downstream of Sardar Sarovar Dam, 1993). After a detailed review of the hydrological regime, which determines the majority of downstream effects, the Wallingford report found that the downstream reach was already heavily populated with a few ecologically rich habitats. It concluded that social impacts are likely to be more significant than ecological impacts. Possible adverse impacts were expected after a considerable period of time. These could be minimised with timely monitoring and appropriate mitigation measures. Its recommendation broadly supported "the programme of studies and actions already proposed."

Wallingford also prepared a comprehensive three-volume report on the impact of agro-chemical pollution due to SSP (Wallingford, 1995). This study identified the current situation of agro-chemical use in the existing SSP area and also in the already functioning Mahi Command scheme. It had done an extensive review of recent literature on chemical pollution in surface and ground waters. This study had assessed the likely impact of agro-chemical use and recommended methods for monitoring and mitigation of its adverse effects.

Wallingford, in its desk review of EIA studies for the command area of SSP, had noted that Gujarat had carried out a large number of studies both on technical and socio-economic issues. It considered the data collection phase as almost over though results of recently commissioned studies were yet to come. It also noted that issues of evolving appropriate monitoring and management were also being addressed on the basis of studies conducted. It concluded, on the basis of available EIA studies for the command area, that there was little justification for not proceeding towards completion of the project (Wallingford, 1993).

While granting environmental clearance to SSP in 1987, MoEF-GoI had stipulated eight conditions which were duly accepted by GoG. Required conditions for catchment area treatment and compensatory afforestation had been fulfilled. Necessary studies on command area, flora, fauna, wildlife and carrying capacity, seismicity, health aspects and archaeological and anthropological aspects had been carried out and required work plans had been prepared. NCA and its environmental sub-groups were regularly monitoring their progress.

It was for the first time in the history of irrigation project planning in India that all the required environmental impact studies had been carried out. All major environmental protection measures suggested by MoEF-GoI had been addressed. Various environmental impact assessment studies had suggested the opportunities available for improving the environment. They also indicated the major environmental challenges posed by the project. Appropriate policy measures had been taken and were also expected in the future for maximising the potential benefits and for minimising likely adverse impacts of this major national project.

VI. Concluding Observations

To begin with, Gujarat had assigned a very high priority to the Dam, Power House and Major Canal constructions. This priority was legitimate and it resulted in several engineering marvels of world class construction. Priority for R&R was also well conceived. Gujarat not only provided one of the most liberal R&R packages, it also ensured its effective implementation. In view of the conditions for environmental clearance laid down by GoI, priority had to be assigned to environmental studies and work programmes where again Gujarat achieved remarkable success. Thus, SSP has already crossed many difficult hurdles to establish itself as a model project.

What now remains to be accomplished relates to water management and command area development. If SSP succeeds in this acid test, it can rightly establish its claim as a model project both within and outside India. SSNNL recently constituted a high level expert committee to make recommendations for speedy execution of command area development programme (SSNNL, 2010). As the major problems of funding and land acquisition have been resolved, GoG has undertaken this work on a war footing. Its early completion will be in the best interest of Gujarat and the country.

Even though SSP has been the most studied irrigation project in India, it was destined to be in the vortex of controversies. It encountered numerous road blocks on its journey over the past five decades. If the first decade, of the 1960s was lost in interstate disputes, the second, of the 1970s was claimed by the proceedings of NWDT. The NWDT award was binding on everyone and, therefore, speedy implementation of the project was expected. The decade of 1980s was claimed by environmental clearance and the anti-dam campaign. In the 1990s, the Supreme Court put a ban on dam construction (1992) which was lifted only in the year 2000. The anti-dam lobby, national and international, was always trying, by fair means or foul, to stall the project. Some of the participating states had their own reservations and were reluctant to contribute their share of project funding. This long-drawn out battle was fought by Gujarat with patience, courage and determination.

Gujarat has thus fought a relentless battle spread over five decades to overcome the innumerable challenges posed by this mega-project. This marathon effort was spearheaded by a determined political leadership cutting across party lines. It was ably supported by irrigation engineers, bureaucrats, professional experts, and NGOs and fully supported by the people of Gujarat.

Appendix I

Members of the Narmada Planning Group (NPG), 1981

Y.K. Alagh
Executive Vice Chairman
(Director, Sardar Patel Institute of Economic and Social Research, Ahmedabad)

S.R. Hashim, Full-time Executive Member (for two years)
(Professor, MSU)
B.R. Deolalikar (Member, State Planning Board)
S.S. Mehta (Director, Perspective Planning, GAD)
Nitin Patel, Part-time Member (served for five years)
(Professor, IIM, Ahmedabad)
B.N. Vishwanath, Part-time Member (served for five years)
(Professor, School of Planning, Ahmedabad)
D.T. Buch
(Iirrigation Engineer)
Gunvant Patel, Adviser
(Former Director of Agriculture, GoG)
P.B. Buch, Adviser
(Former Director, Bureau of Economics and Statistics, GoG)
S.N. Joshi, Agriculture Expert
Faldu (Agriculture Expert)
V.M. Patel (Statistics)

8

Financing the Sardar Sarovar Project: Challenges and Solutions

I.P. Gautam

Introduction

The Sardar Sarovar (Narmada) Project (SSP) is one of the key irrigation and hydro-power projects of independent India with potentially far reaching implications in terms of improvement in agricultural yield, crop diversification, livelihood support and regional growth impetus. The project principally involves construction of the Sardar Sarovar Dam at Kevadia near Bharuch in Central Gujarat on the river Narmada and the branch network of canals for water distribution. It also includes hydro-power generation at the canal head and river head power stations. Water from the project is slated to be shared by the states of Gujarat, Maharashtra, Rajasthan and MP according to a pre-decided formula.

Geographically the dam site falls in Gujarat, and the complete responsibility for the construction of the dam, the power project and all associated infrastructure including canal network in Gujarat has been taken up by the Government of Gujarat (GoG) through its wholly owned company Sardar Sarovar Narmada Nigam Ltd. (SSNNL). Additionally the Sardar Sarovar Rehabilitation Agency has taken up the task of Rehabilitation and Resettlement (R&R) of Project Affected Persons (PAPs) successfully over the last 20 years.

Project Background

The history of the project dates back to pre-independence times. The potential of harnessing water of Narmada was recognised in 1945 when Government of India (GoI) constituted a committee under the chairmanship of A.N. Khosla for planning and harnessing of the river waters. In 1946, the then Governments of Central Provinces and Berar and of Bombay requested the Central Waterways Irrigation and Navigation Commission to investigate the Narmada river system for flood control, irrigation, power and navigation projects. Such studies were conducted and revealed excellent potential storage sites on the river and its tributaries. Subsequent to this, a project report was prepared whereby a dam was planned near Gora village, named as Broach Irrigation Project, with full reservoir level of 160 feet (49 metres) to irrigate about 400,000 hectares of command area. Several debates and discussions on the subject followed.

Meanwhile on 1st May, 1960 the State of Gujarat was formed. Almost at the same time, the Broach Irrigation Project Stage-I was accepted by the Planning Commission in August 1960. However, during the survey of river and its command area, GoI concluded that a much larger area could be commanded with a high level canal off-take at 300 ft. (91 m), and indicated that it was possible to store more water if a dam of greater height was constructed.

Accordingly, the dam site was shifted and detailed engineering and geological investigations were conducted. However, since Narmada is an inter-state river, issues arose, which could not be resolved amicably despite repeated negotiations and dialogues. As an outcome thereof, GoG submitted a petition to GoI in 1968 for appointment of a Tribunal under the Inter State Water Disputes Act, 1956.

Acting upon the plea by Gujarat, GoI appointed a Tribunal in 1969 for adjudicating the dispute over the Narmada water. The Tribunal was headed by a sitting Judge of the Supreme Court and was assisted by expert assessors in the field of agriculture, hydrology and engineering. They examined the detailed claims of the States of Madhya Pradesh (MP), Maharashtra, Gujarat and

Rajasthan, investigated the matter and finally adjudicated the dispute and gave a final decision in December 1979. According to the decision Gujarat was to receive only 32 per cent of the total water and only 16 per cent of the total power production from the project. Despite this, given its precarious water situation and its traditional development appetite, Gujarat was the most willing and proactive member of this group. This aspect is commented upon later in detail.

The Tribunal ordered the setting up of machinery to implement its decision. GoI accordingly set-up the Narmada Control Authority (NCA), Sardar Sarovar Construction Advisory Committee and Review Committee of Narmada Control Authority to perform various functions to implement the decision of the Tribunal. GoG prepared the project report and detailed estimate of Sardar Sarovar Project amounting to Rs. 64.06 billion at 1986-87 price levels which was approved by the Planning Commission.

Financing Arrangement

The initial financing plan for the project was based around the aid from the World Bank (WB) and the Overseas Economic Cooperation Fund (OECF), Japan. GoG prepared the Sardar Sarovar Development Plan (1983) through the Narmada Planning Group and presented it to WB for appraisal. WB examined the designs of the canal system, various parameters of dam design and related aspects.

The World Bank

WB agreed to provide credit assistance totalling US $450 million in 1985 for certain components of works included in Stage I of the project. Out of this, $300 million was for dam and civil works of power houses while US $50 million was for the main canal up to the Mahi River and part of the distribution system under this segment of the canal.

Around the early 1990s, enormous pressure was built up on WB by a group of NGOs and individuals who were opposing the dam. The then WB President, Barber Conable decided to appoint an Independent Review Mission (IRM) to review SSP under the pressure of NGOs and those opposing the Dam. The IRM started its work in September 1991. It equated the Narmada Valley R&R with human rights and cited ILO Convention 107, adopted in 1957 and ratified by India. Finally, GoI informed WB in March 1993 that it was withdrawing SSP from the Bank's lending portfolio, giving up an unspent US $182 million. It instead decided to complete the project without external assistance and agreed to provide additional financial assistance to the project in lieu of the foregone amount of $182 million.

OECF agreed to provide credit of 28.5 billion Yen, equivalent to Rs. 1.5 billion at the exchange rate prevailing in 1986, for electrical components of the river bed power house. SSP did not avail of credit assistance to the extent of US $182 million from WB and almost the entire credit from OECF owing to the conditions mentioned below. This brought into sharp focus the need to raise funds from alternative sources.

Key Decisions of the Narmada Water Tribunal Award

(a) Allocation of Narmada Water at 75 per cent dependable yield:

State	*Water Allocation (MAF)*
Madhya Pradesh	18.25
Gujarat	9.00
Maharashtra	0.25
Rajasthan	0.50
Total	28.00

(b) Height of Sardar Sarovar Dam

(i) Full Reservoir Level	455 feet (138.68 m)
(ii) Maximum Water Level	460 feet (140.21 m)

(c) Full supply level of Narmada Main Canal 300 feet (91.44 m)

(d) Allocation of Hydro power generated at Sardar Sarovar Dam

(i) Madhya Pradesh	57 per cent
(ii) Maharashtra	27 per cent
(iii) Gujarat	16 per cent

Challenges before Project Authorities/State Government for Financing the Project and the Solutions Worked Out

For Financing the Project and Solutions

Challenges

1. After WB withdrawal, financing for construction of dam, power house and canal network as per NWDT Award.
2. Financing and completing R&R for PAFs of Gujarat within Gujarat.
3. Completion of R&R of PAFs of Maharashtra and MP in their states.
4. Compliance of environment issues and various directions of NCA to comply with the condition of NWDT Award and financing.
5. Neutralising organised propaganda against SSP in India and abroad.

Once the Hon'ble Supreme Court had stayed the construction of the dam in April 1994, complying with NWDT Award and financing the project became more difficult.

Responses

Following the WB withdrawal from the project, its financing was a major challenge before SSNNL, GoG and also for the partner states of MP, Maharashtra and Rajasthan. GoG and the project authorities had limited options to finance this kind of multidisciplinary mega-project where the construction of the dam was at a primary stage; main canal work had started, construction of power-house had just started and due to pressure from a few NGOs there was no active participation of partner States.

In this scenario GoG through its SPV-SSNNL—played a lead role in financing the project. For this it came up with some unique innovative mechanisms:

1. It initiated a public deposit scheme called **Srinidhi Fixed Deposit Scheme** for a period of one year, two years and three years to ensure liquidity for the project

2. **Market borrowing** through bond issues: SSNNL issued the first bond of its kind, based on a tripartite agreement in 1993. That was a public issue of Deep Discount Bonds (DDB).
3. After an overwhelming response to the DDB public issue and assessing its servicing cost, it was decided to **approach the capital market on private placement basis.** This was easy to service but difficult to mobilise.
4. By the time the private placement bond was designed the Hon'ble Supreme Court stayed the construction of dam in 1994. In view of the stay the whole **project financing programme was redesigned** in consultation with GoI and the partner states.
   - (i) Due to the limit on the government guarantee, GoG was not in a position to guarantee loans/borrowings. Therefore, it increased its budgetary allocation for the canal network.
   - (ii) GoI made some allocation for command area development under the centrally sponsored Accelerated Irrigation Benefit Programme (AIBP).
   - (iii) SSNNL took the following initiatives:
     - (a) Strengthened the *Srinidhi Fixed Deposit Scheme*, and the entire state government machinery of revenue, panchayat and municipal administration was involved in mobilising funds under it.
     - (b) Awards were instituted for Best Collector, Best DDO, Best TDO and Best Mamlatdar on the basis of fund mobilisation for SSNNL, and for the first few years (1993-1996) the major source of financing was the Srinidhi FD Scheme.
     - (c) Boards, Corporations, Cooperative Banks, Cooperative Societies, Sugar Mills, Sugar Federations and Farmer Societies were requested to deposit most of their surplus funds in the Srinidhi FD Scheme. It can be asserted that from 1993 to 1996 SSP was financed by people of Gujarat. Both institutions and individuals exceeded expectations and took up the challenge of supporting the project.
   - (iv) SSNNL institutionalised its bond issue was were issuing bonds every year after 1995 on private placement basis.

> Due to the marketing strategy adopted this bond became so popular and marketable that most cooperative banks, commercial banks, private banks and noteworthy individuals from all over India subscribed to it. The investor base in SSNNL's market borrowing programme was countrywide.

Owing to the Supreme Court stay, construction of the dam was stopped and the future was very bleak. The credit rating of GoG and loan servicing capacity of SSNL were carefully marketed all over India. The investors included J&K Cooperative Bank, Andaman and Nicobar Cooperative Bank, Sahara India, Peerless India, all nationalised banks, financial institutions like ICICI, IDBI, LIC, Central Government Public Sector Undertakings like FCI, Provident Funds and all major Port Trusts of the Central Government.

1. Srinidhi Public Deposits

Over 600,000 investors have invested an aggregate of Rs. 25 billion under this scheme so far. The annual collections from it provided important liquidity to the SSNNL due to its cyclic and continuous inflow. This can be seen from the figures of annual collections from the scheme over the years.

Year	*Rs. in crores*
1993-94	161.13
1994-95	37.17
1995-96	32.00
1996-97	50.00
1997-98	95.00
1998-99	103.00
1999-2000	175.00
2000-01	370.00
2001-02	492.78
2002-03	573.91
2003-04	408.33
2004-05	400.94
2005-06	215.22
2006-07	138.11
2007-08	133.73
2008-09	157.61
2009-10	1249.68
2010-11 up to January 2011	460.64

The significantly higher collection in 2009-10 is due to a renewed drive to collect resources through this means.

2. Market Borrowing

As discussed earlier, SSP was one of the key projects for irrigation and development in Gujarat and it became closely associated with the state's identity. The non-availability of accept WB and OECF funding led to considerable thinking and debate within the Government and sections of society about alternative ways of funding the project. This led to the idea of public participation as a means of financing the project.

The idea of issuing bonds was innovative at that time for a number of reasons:

1. Financing water infrastructure projects through public participation was *per se* unique. Few infrastructure projects had been funded from outside budgetary resources, and almost none had borrowed through public issue of securities.
2. Before the first bond issue in 1993, few projects of this kind in India had borrowed through open market issue of bonds. The bond markets in India at that time primarily consisted of Government issues (Government Securities or G-Secs for short). The buyers and sellers of these bonds were mostly banks and some institutional investors who invested to meet statutory requirements. These markets were not deep and were quite lacking in liquidity.
3. The structuring of the bonds was unique. Their repayment was through a Structured Payment Mechanism under a tripartite agreement between GoG, SSNNL and the Trustees to the issue. There was also a special escrow account for servicing them. This was a source of great comfort for the bondholders.
4. Further, while there were many corporate issues, these were limited to AAA rated bonds whose creditworthiness was considered close to sovereign levels. None of them was for long gestation infrastructure projects where returns were highly back-ended and repayment was promised not from project cash flows but from budgetary allocations.

3. Deep Discount Bond

SSNNL attempted the first bond issue in 1993 through the issue of 20-year maturity Deep Discount Bonds. The face value of each bond was kept low to facilitate wide public participation. The return on the bond as represented by the amount promised on maternity was attractive by today's standards but it must be remembered that money markets up to the late nineties were experiencing a high credit crunch and money was costly and difficult to come by.

The public issue of the Bonds for an aggregate amount of Rs. 256.90 crore and Non Convertible Bonds of Rs. 118 crore was oversubscribed, with the overwhelming response of more than seven lakh investors from all over the country. A detailed analysis of investors revealed that about 50 per cent of the investment came from the states of Gujarat and Maharashtra, mostly from Gujarati community. But investors from all over India supported the initiative.

4. Private Placement

The successes of Deep Discount Bonds in public financing provided further encouragement for similar bonds later but by way of private placements. SSNNL thereafter floated several Private Placements of Bonds as detailed below:

Sl. No.	*Year*	*Amount raised through bond issue (Rs. crores)*
1.	1995	302.00
2.	1997	547.00
3.	1999	2046.00
4.	2000	269.00
5.	2001	700.00
6.	2002	1100.00
7.	2003	300.00
8.	2004	1250.00
	Total	6514.00

Thus, the withdrawal of WB and OECF from financing SSP was converted into an opportunity to enlist public participation, by way of raising finances from the Indian people. Leading co-operative banks, PF Trusts, Charitable Trusts, Port Trusts, Insurance Companies, Banks and financial institutions were major investors in the private placement. Particular mention must be made of the contribution by cooperative banks of the state, which participated enthusiastically in the bond issues in keeping with the spirit of the development of the state. Their members belonged to rural and semi-rural areas, and could directly appreciate the benefits of the project.

Another key feature of the bond issues was the participation of a wide range of institutions from across the country. It may be mentioned at this point that the private placement method was adopted subsequent to the first bond issue not only to limit issuing costs but also to directly tap the treasuries of institutions with deep pockets. This method benefited the treasuries of religious and charitable trusts, Provident Funds of large institutions such as the State Bank of India, ONGC, the oil companies, as also treasuries of organisations that were traditionally endowed with such funds due to the nature of their business, such as insurance companies, port trusts and banks. Hitherto these treasuries had few diversified investment options and their principal investments were in Government securities of the kind mentioned earlier. The SSNNL bonds were structured to allow them the following benefits:

Rated paper by CRISIL/CARE with structural obligations.
Relatively attractive returns.
Option of Put, meaning option to encash and exit from the bonds through redemption at pre-decided intervals.
Option of liquidity through sale as these bonds were listed thoughtfully by the SSNNL on the debt market segment of the National Stock Exchange despite trade volumes through that route being thin in this segment in those days. The NSE debt market segment picked up later benefiting many investors.

5. Term Loans from Banks

The Honourable Supreme Court had stayed the construction of the dam, but work on the main canal, river bridges, approach bridges and distributaries continued, and also huge structures crossing seven to eight rivers from Kevadia to the Saurashtra Branch Canal. Therefore, fund requirement increased during this period and SSNNL took short-term loans from the nationalised banks from time to time. Most of the Nationalised Banks in the country have supported this Project with short- and medium-term loans. In 2004-05 SSNNL approached Power Finance Corporation (PFC) of GoG for finance to construct the 1200 MW River Bed Power House. PFC gave Rs. 1 billion term loan for the purpose.

6. Suppliers' Credit

Since the Hydro Power Project was partly financed by OECF, an order was placed with Sumitomo, Toshiba and Hitachi, Japan to construct/manufacture six turbines of 200 MW each which had been fabricated by these companies before 1995-96. But due to the Supreme Court's stay it was difficult for SSNNL to take delivery of them in absence of the finances earlier sanctioned by OECF. In 1997 the Japan Government represented to Ministry of Finance seeking resolution of the issue of supply of and payment for these turbines. The then Finance Secretary, Montek Singh. Ahluwalia agreed in principle to a sovereign guarantee if SSNNL was taking supplier credit from Japan and getting delivery of these Turbines. Actually Gujarat's share in the hydro-power from SSP was only 16 per cent and GoG was supposed to spend only 16 per cent of the capital cost. However, in the interest of project completion and generation of green energy, GoG gave a guarantee of about Rs. 10 billion for supplier's credit from Sumitomo and took delivery of the turbines. It was a historic and exemplary decision, where other participating states were not very keen to support the loan, but in the interest of the project, GoG borrowed the full Rs. 10 billion for the turbines when the state's power share was only 16 per cent.

7. Financing R&R Work in MP and Maharashtra

In 2001-02 GoG was facing a major financial crunch in the aftermath of the earthquake. At the same time, as per the Supreme Court decision for further raising the dam height, R&R work in Gujarat, Maharashtra and MP had to be completed before NCA could take a decision on the dam height. All three states had constituted Grievance Redressal Authorities (GRA) and R&R was to be done to the satisfaction of the respective GRA. When the NCA team met the concerned officers of MP and Maharashtra, they pleaded their inability to complete R&R work in the given time frame due to the financial crunch faced by their States. The Chief Minister of Gujarat had spoken and held several meetings with concerned Chief Ministers. Chief Secretary level meetings also failed to bring any fruitful results. Both the States were supposed to pay their share of project expenditure which was running into billions of rupees. Reminders were sent to both the Governments from time to time but the money could not be recovered. It was a big issue how to raise the dam height unless the two partner states completed their share of rehabilitation. At this juncture GoG took another exemplary decision to finance the R&R work of MP and Maharashtra. Billions of rupees in advances were released in stages to both the states to complete their part of R&R to fulfil the NCA's conditions before due dates. This was a unique decision. GoG had provided finance to defaulting state governments for faster completion of this Project.

Conclusions

In conclusion, the following points stand out in the solutions used to meet the challenges of financing the Sardar Sarovar Project:

1. The people of the country in general, and particularly the people of Gujarat, participated widely in financing the project. In this respect it became a pioneering project where large scale goodwill for it helped raise finances.
2. While the benefits from the project to Gujarat in the Power sector were smaller than they were to the other two candidate states, this did not become a limiting factor for the state government to find alternative and innovative ways to fund the project.
3. During the majority of the bond issues, the increase in height of the dam was restricted by the Supreme Court while it heard those who opposed the dam. (The Court allowed the height to be increased in stages after it was satisfied about the R&R

package being given by the state to the PAPs). However, this factor did not dampen the spirit of the public or the institutional investors who trusted the merits of the project and the commitment of SSNNL and GoG.

4. None of the partner states other than Gujarat participated in raising funds for the project. The responsibility to raise and repay funds was taken up solely by Gujarat. For R&R work GoG made advance payments to partner states in the interest of the project. This was a great challenge, particularly when funding was through open market borrowings and therefore, was open to risk scrutiny by sophisticated institutional investors.
5. GoG also used its budgetary resources to convey Narmada water to a thirsty Saurashtra through a pipeline in the drought year of 1999. GWSSB had laid hundreds of kilometres of drinking water pipeline from Kheda District to Junagadh, and also from Surendranagar to Morbi. This water was supplied to cities like Bhavnagar, Amreli, Rajkot, Junagadh and Morbi, which helped the people at a critical time and created goodwill for the project.
6. Maximum credit goes to People of Gujarat, NGOs, the Cooperative Sector, State Public Sector Undertakings, Religious and Charitable Institutions, Educational Institutions, Organisations of Farmers from various parts of the state and the Gujarati media for rising to the occasion not only to support the State Government and Project Authorities in providing alternative sources of funding but also in helping to build a positive opinion in favour of the Sardar Sarovar Project. This helped the Project Authority in raising resources during critical period.

9

Environmental Upgradation through Sardar Sarovar Project

Sanat A. Chavan

Development has been equated with improvement in the quality of life of people. However, most development measures affect our natural wealth—air, soil, water, forests, minerals and our wild ecosystem. Since these resources are exhaustible, their depletion leads to serious consequences on our environment. Environmental planning of a project, therefore, involves a series of measures to ensure that sustainable development takes place without destroying our environment. Therefore, the importance of upgrading the environment even as development takes place has now assumed importance.

Ministry of Environment and Forests, Government of India (MoEF-GoI), armed with the powerful Forest Conservation Act, 1980, subjected the Sardar Sarovar Project (SSP) to exacting environmental conditions before accepting it. Thus it became the first River Valley Project in India that is complying with stringent environmental standards.

Environmental Impact Assessment

It is now well recognised that all development activities impinge on the natural environment and create stress. The efficacy of large dams instead of small dams is debatable. However, the challenge lies in detailed and correct environment assessment study of all the positive and negative impacts so that the negative impacts can be mitigated for regaining or maintaining the environmental balance.

Some of the ecological parameters for Environmental Impact Assessment (EIA) that should be studied as per the guidelines for EIA of river valley projects, issued by MoEF-GoI in 1985 are as under:

(i) *Location Aspects*—The site should be chosen keeping the following in view:

Short and Long Term impacts on population/human settlements in the inundated and watershed areas;
Impact on flora and fauna (wildlife) in the vicinity;
Impact on wildlife (including birds) breeding area/feeding area/migratory route;
Impact on National Parks and Sanctuaries—both existing and potential.
Impact on sites and monuments of historical/cultural and religious significance;
Impact on forests, agriculture; fishery, recreation and tourism etc.

(ii) *Physical Aspects*—Impoundment alters surface water patterns with great impact on underground aquifers and their recharge. Major aspects to be considered include:

Landslides on the periphery of the dam;
Sedimentation expected in the reservoir—identification of critical areas susceptible to erosion and methods of possible treatment;
Groundwater recharge or other groundwater changes;
Expected water quality (salinity) changes over time and their effect on the river ecosystem, both at the point of impoundment and downstream;
Land-use patterns and practices in the vicinity of the water spread that would affect aquatic growth patterns;
Potential seismic impact of reservoir loading;
Favourable aspects of impoundment on project area.

(iii) *Resource Linkage Aspects* include:

Resource trade-off, such as loss of optional land uses due to impoundment—mineral deposit loss, forest reserve diminution, monuments inundated, recreational facilities lost, dislocation of existing settlements etc.;
Compatibility of dam's creation and operation with present or planned development of the region;

Effect on resident and migrating fish and other aquatic life and assessment of new fishing potential.

(iv) *Socio-cultural Aspects* to be considered before relocation of the people include:

Population relocation requirements in inundated as well as watershed area;

Identification of educational and vocational training programmes to be imparted to the affected population so that they can cope better with the new life style;

Resettlement area planning for housing and other amenities of community life.

(v) *Public Health Aspects:*

New health problems or vector problems that may arise due to changes in water velocities, temperature, quality or other physical change factors caused by water impoundment;

Facilities for migrant construction workers and immigrant influx;

Measures to control contamination of surface and groundwater due to pesticides and fertilisers.

(vi) *Cost Benefits Analysis:* The cost of proposed remedial and mitigation measures, if any, to protect environment must be included in the project costs. Mitigation measures include:

Compensatory afforestation;

Restoration of land in construction areas by filling, grading etc. to prevent further erosion;

Control of aquatic weeds in submerged areas to provide improved habitat for aquatic life;

Measures to salvage/rehabilitate any rare or endangered species of fauna found in the affected area;

Measures to salvage and relocate monuments from inundated zones;

Enforcement of anti-poaching laws;

Measures to prevent forest fires, overgrazing etc.;

Establishment of fuel depots to meet fuel requirements of labour force to prevent indiscriminate felling of trees;

Public health measures to control spread of water and soil-borne diseases.

The EIA guidelines also suggest that field survey and studies should be undertaken to create the environmental database and technical and administrative measures to be in place to effectively monitor the observance of suggested safeguards and mitigation measures.

The Narmada Water Dispute Tribunal (NWDT) constituted under Inter State Water Disputes Act gave its award for building the Sardar Sarovar Dam in December 1979. The project was one of the first in India to incorporate environmental impact assessment in the planning stage itself, even before the guidelines for such study were prepared by MoEF-GOI. A benchmark study, 'Preliminary impact assessment of Sardar Sarovar' by M.S. University of Baroda was carried out in 1983. This was followed by a number of studies relating to all aspects of environment such as health, downstream impacts on fisheries, problems related to water logging, salinity, seismicity, impact on flora and fauna and also the socio-economic analysis of the Project Affected Persons. Besides these specific studies there was a plethora of command area studies carried out on various aspects. For example, the Narmada Planning Group, once headed by the noted economist Y.K. Alagh, has carried out a total of 41 of studies grouped under Technical Studies (18), Socio-economic Studies (10), Agricultural Studies (6) and, Environmental Studies (7). There were also a number of studies carried out by the Government, non-government and independent bodies making it one of the most exhaustively researched projects in India.

MoEF-GoI tested its guidelines for environment impact assessment by subjecting SSP to one of the most exacting sets of conditions. After satisfying itself about the sustainability of Sardar Sarovar Dam, it granted clearance in 1987 under the Forest Conservation Act (FCA), 1980, with a condition that all amelioration activities would be carried out *pari passu* with the construction of the project. The Planning Commission reiterated these conditions while accepting the project.

One of the most glaring impacts of a river valley project is the loss of forest land that comes under the vicinity of the large man made reservoir. Tables 1 and 2 give an indication of the forest land diverted for SSP and the efforts of various states to carry out compensatory afforestation to mitigate that impact as per the provisions of FCA, 1980. In Gujarat it was estimated that approximately 828,000 trees would be submerged. Gujarat, therefore, has already planted more than 60 million saplings, i.e. approximately 77 plants for every tree clear felled. The species planted show a wide diversity suitable to the agro-climatic conditions of the region. The plantation programme in Gujarat included canal bank plantations which have a vast potential of an estimated 5300 hectares.

Environmental Upgrading (till 2002): The greening of the dam-site is an integral part of the project design. The dam site abounds in natural beauty, which was threatened to some extent because of human disturbances due to project activity. A survey conducted in Kevadia and the surrounding project impact areas showed that the forest villages found more than 60 uses of the various species planted around them. The ultimate aim was to convert the area into an eco-tourism zone so that both man and nature could live in harmony. A study of the environmental impact of compensatory afforestation in Kachchh has given encouraging results. Before plantation, the area was degraded with poor unpalatable grass species. These species, such as *Eragrostis* and *Aristida* have been replaced by better species like *Dicanthium*, *Cenchrus sp.* The chinkara population has increased three to four-fold. Birds like Great Indian Bustard, Houbara Bustard, Larks, Pipits and Chats have also significantly increased. The plantations are acting as windbreaks and providing employment and fodder to the local populace.

Table 1: State-wise Statistics of Specialised Land use

State	*Area of forest diverted for SSP (ha)*	*Area of non-forest land to be afforested (ha)*	*Achievement (ha)*	*% task*
Government of Gujarat	4523	4650	4650	100
Government of Maharashtra (i) Owing to submergence	6488	6488	6326	97.5
(ii) Owing to R & R	4200	4200	3583	85.32
Government of Madhya Pradesh	2732	2190	X	X
Total	17943	17528	X	X

X—Not reported

Source: Status Report on Environment, Narmada Control Authority, June 1997.

Table 2: State-wise Statistics of Degraded Forest to be Afforested

State	*Area of degrade forest to be afforested (ha.)*	*Achievement (ha)*	*% of task*
Government of Gujarat	9300	9300	100
Government of Maharashtra	12980	12977	100
Government of Madhya Pradesh	6550	X	X
Total	28830	X	X

X—Not reported
Source: Status Report on Environment, Narmada Control Authority, June 1977.

The condition of the catchment area of any dam is an indicator of the health of the dam. The committee of experts (Diwan Committee) set up by Ministry of Agriculture suggested various measures required for protection of the catchment area of the Sardar Sarovar Dam. MoEF-GoI issued a directive in June 1992 that for SSP, the project would bear the cost of treatment of all critically degraded sub-watersheds directly draining into the reservoir. The watersheds were identified with the help of All India Soil and Land User Survey (AISLUS) the respective states took up for treatment plans for implementation (Table 3). In Gujarat, the entire catchment area was classified, based on stock mapping survey by the Forest Department and also by photo interpretation of satellite imageries obtained from ISRO with LISS-II censors. A preliminary interpretation of catchment area by satellite imagery carried out by GEER foundation, Gandhinagar, shows a distinct increase in the density-wise area of catchment of SSP between 1985 and 1994.

There are encouraging reports that more concentration of birds is now seen in these areas including canal bank plantation areas. Significant among them are Golden-back woodpecker, Golden oriole, Brain-fever bird, Green Bee Eaters, Marsh Harriers etc.

Wildlife Affected Areas (Sanctuaries and National Parks)

The wildlife of the affected areas forms an important component of the Environmental Impact Area Study. Under the influence of the Sardar Sarovar Project, the area of Shoolpaneshwar Sanctuary was

Table 3: Programme Target and Achievement of Various States

State	*Total catchment (ha)*	*Directly draining high and very high areas planned for treatment (ha)*	*Forest Area (ha)*	*Non-forest area (ha)*	*Achievement*	
					Forest Area	*Non-forest area*
Government of Gujarat	30,230	29157	27204	1953	27042	1953
Government of Maharashtra	163610	22768	20000	2768	1875	3011
Government of MP	2248600	125725	51930	73,795	X	X
Total	2,442,440	177650	99134	76,516	X	X

X—MP has reported treatment of 52,571 ha. area which includes catchment in non-forest.

Source: Status Report on Environment, Narmada Control Authority, June 1997.

increased. Two studies were undertaken simultaneously by M.S. University of Baroda team lead by Prof. S.D. Sabnis and other study by this author and his team. In the author's study, some rare and endangered fauna and avifauna were sighted and recorded. These include the Dangs giant squirrel (first recorded in the Rajpipla forests), Rusty spotted cat, Leopard cat, Sambar, Hanuman langur, Rhesus, Sloth bear, Jungle cat, Civet cat, Porcupine, Jackal and Fox, Hyena and Mongoose. Most significant of these was the Dangs giant squirrel (*Ratufa indica dealbeta*), recorded in the Shoolpaneshwar Sanctuary area, which has been declared extinct in whole Dangs forest area for over 60 years.

The avifauna sighted and recorded included rare, endangered as well as beautiful birds such as Red and Grey jungle fowl, Spur fowl, Pea fowl, Racket tailed drongo, Golden and Black headed oriole, several species of woodpecker (including Golden-backed), large number of owl species (including Mottled-wood owl, Brown fishing owl) and a very rich collection of raptors (birds of prey). But the most important was the record of two bird species, new for Gujarat state earlier recorded by the late Dr. Salim Ali in the Eastern Himalayas—Hodgeson's frog-mouth and Streaked spider-hunter.

It was mainly due to these studies and findings of rich fauna and avifauna that the sanctuary area was increased from 150 km^2 to 607 km^2 in the catchment area of the Narmada. A habitat improvement programme was undertaken in the sanctuary, based on a detailed action plan prepared by the Environmental Cell of SSNNL. An expert group was also constituted to undertake detailed studies on improving the habitat of the sanctuaries in Narmada Command Zone, i.e. Shoolpaneshwar Sanctuary near Rajpipla, Blackbuck-Velavadar National Park, Nalsarovar Bird Sanctuary near Ahmedabad, and Wild Ass Sanctuary in the Little Rann of Kachchh.

It is expected that no serious adverse impact is likely to fishing and fishermen, but there will be some changes in the variety of fish and the method of fishing, which may reflect on the livelihood of fishermen. Detailed action plans have been prepared to develop the dam reservoir as a fishing gene pool. Every year fingerlings are released into the reservoir by GoG through the Environment Cell of SSNNL to stock the reservoir with edible variety of fish. Adequate measures were taken to train habitats for sustainable fishing to fish in the periphery of the dam reservoir and mangrove plantations were taken up in estuaries to promote fishing.

Other Aspects of Environment

Health: When a mega project comes into existence, various factors like creation of a man-made reservoir, long canal system and influx of human population around the project site can lead to health problems. The EIA identified the baseline indicators for vector transmitted diseases, including malaria, filariasis and schistosomiasis, and recommends alternative solutions for implementation. Initially, a study was taken up for schistosomiasis in the project area and it was inferred that it did not exist. A number of studies have been carried out on the health profile of the villages in the affected areas. A separate unit of preventive health organisation at Kevadiya was also set up. Intensive monitoring for malaria was regularly done in the project area.

Archaeology: A survey was carried out in Gujarat in respect of archaeological monuments affected due to submergence. Two ancient temples on the Narmada river banks in Shoolpaneshwar and Hampheshwar have been shifted to newer sites above the submergence area. In MP, seven temples have been relocated.

Seismicity: Studies on reservoir induced seismicity (RIS) and rim stability were carried out by Geological Survey of India, Central Water and Power Research Station, Roorkee and World Bank consultants. Based on the recommendations emanating from these studies, various modifications in the dam design were carried out.

There were other aspects of the Environment Impact Assessment of the River Valley Project, like Ground Water Modelling, dealing with rise of ground water over time as the canal water starts following in; understanding the drainage pattern; distribution of water etc. which have been addressed systematically by SSP. Rehabilitation—a prime environmental concern has also been addressed with sensitivity and empathy.

It is to be stressed that the re-establishment of green cover that would otherwise have been lost in such a massive river valley project will benefit millions of people for generations. The environmental mitigation measures carried out by SSP are unparalleled anywhere in the country. However, the fragility of the ecological balance calls for tough measures for continuous environmental monitoring and the right guidance to the user groups, so that the ills of other similar projects are checked at the most opportune moment.

It is, therefore, imperative to learn from the experiences, and be ever vigilant on the environmental front, for taking of effective actions in the project life cycle. It will be an attempt to give back nature its due.

PART 3

Appraisal for SSP

10

Review of Social Cost-Benefit Analyses for the Sardar Sarovar Project

Ravindra H. Dholakia

1. Introduction

The Sardar Sarovar Project (SSP) on the River Narmada is one of the most debated, most researched and most meticulously planned water projects in the world. Interestingly, both the proponents and the opponents of the project have used the cost-benefit analysis to derive support and comfort for their respective strong positions. The project got caught in long drawn out legal disputes, political controversies, demonstrations and agitations. That it could ultimately emerge from this mire after consuming precious time of about two decades, and reach the stage of getting successfully commissioned during the first decade of the 21st century speaks volumes of the firm commitment and determination on the part of interest groups on both the sides. It has provided an excellent illustration of a very systematic, tactical and strategic negotiation process among several societal and individual stakeholders. Since democratic values were not sacrificed during this process, the final outcome can be viewed in the framework of rational social choice. Thus, if the project could ultimately be commissioned, it would show that it was finally found socially and economically acceptable and viable at least at the end of the struggle. The social cost-benefit analysis (SCBA) of a project is all about judging its economic and social viability upfront by an economic appraisal on *ex-ante* basis. SCBA as an important tool for guiding public policy decisions on selecting or rejecting such critical projects is usually carried out *ex-ante* rather than *ex-post*. Also, such an analysis is invariably done considering the costs and benefits of the project to society as a whole rather than to an exclusive institution or individual entity.

When the initial approval to SSP was given by the tribunal for more specific designing and planning of details, an estimate of costs-benefit ratio (CBR) was required as is the usual prerequisite for all irrigation projects in the country. However, given the size, magnitude and temporal dimension of SSP, the Narmada Planning Group (NPG) decided at the outset to commission a comprehensive economic appraisal of the project, including estimation of its social costs and benefits, by a reputed outside consultant. Accordingly, Tata Economic Consultancy Services (TECS) Limited was assigned the study and it submitted its final report by 1983. The World Bank also carried out in 1985 an independent appraisal (called Staff Appraisal Report or SAR) since it was financing a small part of the cost of the project. Subsequent developments and public opinion (Alvares and Billorey, 1988; NBA, 1990) on the project required the World Bank to revisit the estimates made for the social costs and benefits of SSP to include the likely effects of the project on environment and ecological balance. The World Bank carried out the whole exercise of re-estimating and recalculating the costs and benefits of SSP with in-house expertise in 1990 (see Levenhagen, 1987; Dixon *et al.*, 1989; World Bank, 1990). The opponents of the project (Paranjpye, 1990; NBA, 1990) had several objections and doubts about the cost-benefit exercises done by TECS and the World Bank. These opponents came out with alternative estimates and analysis of the cost and benefits of the SSP in early 1990. Considerable national and international pressure was built against SSP and consequently, the World Bank had to commission an Independent Review under the chairmanship of Bradford Morse in September 1991. Its report, submitted in June 1992, recommended stepping back of the World Bank from the project to re-evaluate the situation and to institute improvements (World Bank, 1995). However, the World Bank believed that 'the establishment of .. measurable benchmarks for addressing the situation was preferable to stepping back as recommended by the Independent Review' (World Bank, 1995). Accordingly, the economic analysis of SSP was again updated by the World Bank for its project completion report (PCR) in 1993, reflecting changes in quantities, prices and phasing besides including all environmental and ecological costs and benefits. In 1995, the Sardar Patel Institute of Economic Research (SPIESR) attempted to resolve the issue by comprehensively estimating social costs and benefits of SSP with methodological comments.

The present review considers all these past estimates critically in terms of concepts, evaluation methodology and breadth of coverage. Since the opponents of SSP (see above) did not make any fresh independent estimates of costs or benefits, but only modified the existing estimates of TECS (1983), they are examined only for the modifications suggested and for interpretation of the results made by them. The next section discusses several distinguishing features of SSP compared to similar projects. Section 3 identifies all different benefits and costs of SSP to society and discusses certain fallacies of including or excluding some items in the previous studies. Section 4 reviews the evaluation methodologies followed in the SCBA studies of SSP. The last section considers all the evidence on the (internal) economic rate of return (ERR) with sensitivity analysis for the SSP appraisal to conclude.

2. Distinguishing Features of SSP

SSP has several special distinguishing features among comparable projects in terms of size, structure and dimensions of land area and population impacted by the project. The following distinguishing features are particularly relevant in the SCBA of the project:[1]

2.1 SSP is a Mega-project by Any Standards

1.87 million ha. total command area; 37,533 ha. of area inundated or submerged by the reservoir; the longest network of canals in the world; second largest concrete gravity dam in the world; and a huge budget exceeding that for all the major and medium irrigation projects of not only the Centre but also of all states put together during India's

1. Many of these features of SSP have been often mentioned without relating them to SCBA of the project or even examining their implications for any such exercise (see Pathak, 1991; Sheth, 1991; SPIESR, 1995).

Seventh Five Year Plan. This is an extremely important feature. It implies that the economic appraisal of such a mega-project cannot be done on a 'mutually exclusive projects' basis, but only on an 'independent project basis,' because effective comparable alternatives to SSP simply do not exist. The critics of SSP often argue in favour of considering a series of small and medium irrigation projects as an alternative to SSP, without fully appreciating that (i) these are not strict substitutes but complementary to SSP; and (ii) the quantity of fresh water harnessed by SSP around the area is simply impossible with any substitute—leave aside small and medium irrigation schemes. Thus, the decision involved in the economic appraisal is of a 'select-reject' type and not of a 'ranking' type. Therefore, the cut-off rate of return has to be very carefully selected for such a mega-project. It has the potential to shift the growth path of the economy. Hence the cut-off rate has to be consistent with the growth path anticipated for the whole economy at the time of appraisal.

2.2 SSP is a Multi-purpose Project

It also generates hydroelectric power, augments drinking water supply to several hundred villages and towns, augments water supply for industrial use, leads to development of fishery, creates tourist spots at multiple points, controls floods in the downstream etc. Several proponents as well as opponents of SSP have ignored this feature of the project and treated it either as an irrigation project or as a power project, which obviously would not do justice to the economic appraisal of the project.

2.3 SSP is a Multi-state Project

Within the federal structure of India, SSP cannot be viewed as a project of Gujarat state alone. It is truly a multi-state project with benefits likely to flow to the whole nation through well defined forward and backward linkages. Similarly, a large part of the costs of SSP in terms of area inundated by the reservoir, displacements of people from the submerged land etc. are borne by states like Madhya Pradesh (MP) and Maharashtra. However, this feature of the project is not fully appreciated by several of its supporters and critics. From the public finance angle, SSP should be ideally treated as a national and not a regional project. However, all the appraisals except those by World Bank (1985, 1990, 1995) have viewed it from a Gujarat State angle. The opponents (NBA) also considered it as the project of Gujarat throughout their agitation.

2.4 SSP Inundates Only 2 per cent Area vis-à-vis the Potential Command Area

Among all comparable large irrigation projects in the world, SSP has the lowest technical ratio of submerged area to potential command area (Joshi, 1991). Moreover, the pre-submergence density of population in the submerged area was certainly not more than that in the potential command area. In fact it was substantially less given that a large part of it was tribal. Thus, the ratio of adversely affected population on account of SSP to the benefitting population would be certainly less than 2 per cent.

2.5 Only Four Villages Out of a Total 244 Affected are Fully Submerged

The rest are only partly submerged: some people would lose their residence but not land, while some would lose land but not residence. Moreover, only some of the land is submerged round the year or for some days/months a year. In most of the area or for the rest of the year, it is available for regular productive use. This raises the possibility of receding waters yielding extra fertility benefits. Since this feature of the project is not sufficiently highlighted, most of the SCBA and appraisals of SSP carried out have not considered such benefits explicitly for serious estimation.

2.6 The Length of the Reservoir is 214 km

The location of the project is ideal for storing a huge quantity of water in the Narmada river valley with minimal effort. Considering 214 km on one bank, it exposes a total of 428 km stretch of land to the reservoir. Fluctuation of water level over the year, according to the extent of rainfall in the catchment area, would give rise to what is known as draw-down agriculture which would cover a large area.

For SSP, the draw-down agricultural benefits are likely to be substantial compared to other similar projects. None of the appraisals of SSP has rigorously attempted to estimate these benefits. World Bank (1990) made only a casual attempt to estimate it, mainly to flag it among the project benefits.

2.7 SSP is the Terminal Project in the Narmada River Valley

On the upsteam, there are several projects on the river and the most critical among them is the Indira Sagar Project (ISP) in MP. It is a single state project and does not share several features with SSP. However, whether it is built or not has an important bearing on the costs and benefits of SSP.

2.8 The Main Outputs of SSP are Water and Power

These have been the two major constraints to the economic and social development of the participating states—Gujarat, Rajasthan, MP and Maharashtra. The valuation of these outputs must fully recognise the overriding nature of these constraints for all these states. The critics and the appraisers of SSP have perhaps not fully appreciated the nature and extent of the constraints. The award by the Narmada Water Dispute Tribunal (NWDT) in 1979 judiciously divided and distributed project outputs among the four states. It is binding on all parties and cannot be changed. Its allocations are as follows:

State	*Share of Narmada Water*		*Share of power % from SSP (Total 1450 MW)*
	Million Acre Feet	*Percentage*	
MP	18.25	65.18	57
Gujarat	9.00	32.14	16
Rajasthan	0.50	1.79	–
Maharashtra	0.25	0.89	27

2.9 SSP has Very High Positive Distributional Impacts

A large proportion (70%) of the potential command area of SSP falls in hitherto drought prone districts. As a result, a high proportion of small and marginal farmers would get the benefit of irrigation on their fields. Moreover, MP getting a sizeable chunk (57%) of the power generated would gain a huge opportunity for industrial and commercial development generating proportional off-farm employment opportunities. MP is economically less developed and has a large population of the poor and the unemployed.

3. Identification of Social Costs and Benefits of SSP

A careful and comprehensive identification of the cost and benefit items of such a mega-project to society as a whole is a crucial step in the project appraisal exercise. There are certain well laid down principles to carry out this exercise (see, for instance, Musgrave and Musgrave, 1989, Ch. 9). These principles of SCBA for a project are:

3.1 The items to be included should represent costs or benefits of the project to the whole of society or a part thereof. It should not consider costs or benefits only from the viewpoint of an institution, individual entity or the government. SCBA of a project is not the cost-benefit analysis for a government. From the long drawn out history of the debate, controversy and struggle on SSP, one has to note that a lack of clarity on this basic principle on the part of some participants led to unnecessarily rigid positions and waste of time in social negotiations.

3.2 The costs and benefits should be strictly incremental and ascribable only to the project. The relevant items should be causally linked to the project. Since the world is dynamic and things change due to several simultaneous stimuli and factors, identification and evaluation of changes exclusively due to the project are challenging both conceptually and practically. It requires abstraction and plausible assumptions for estimation. Conceptually, it requires consideration of scenarios *with* and *without* the project. In project appraisal, it is often erroneous to consider scenarios *before* and *after* the project to identify and evaluate costs and benefits.

3.3 Unlike financial cost-benefit analysis, SCBA does not consider only cash inflows and outflows. It also considers the effects of the project, both direct and indirect, on society. The indirect effects are often not monetised and internalised into the project. They are the genuine externalities which cause the private costs and benefits to differ from the social costs and benefits of the project. For a mega-

at the first round. They would recur in many more rounds. Public investment of such magnitude is bound to have well defined multiplier effects on private consumption, imports, employment and income. It may also induce investments and savings effects in the economy, ultimately affecting the growth path of the economy.

Generally, for small projects, these multiplier and growth effects are not considered within SCBA because if they are analysed within the framework of a 'mutually exclusive projects' category, the alternative projects considered in the analysis would also have similar multiplier and growth effects. However, when appraising a mega-project like SSP, which is analysed within the 'independent projects' framework, both the multiplier and the growth effects must be explicitly considered. The alternative to the project is not to have the project. The estimates of multiplier effects in the economic appraisals of SSP seem to be highly conservative, perhaps because they have used the narrow concept of first round income multiplier based on an input-output matrix. They have not estimated the private consumption and import multipliers. Similarly, they have also not estimated the growth effects of SSP.

The probable justification of not explicitly considering these growth effects of SSP could be that the 'cut-off' rate of return assumed for the project is the one given by the Nitin Desai Committee set up by the Planning Commission. The Committee had recommended 9 per cent (quoted by Paranjpye, 1990: p. 172) as the cut-off real rate of return considering the national targets of growth, investment rate, incremental capital-output ratio (ICOR) etc. for the Sixth and Seventh Plans within the framework of the perspective plan for a longer period. It can be shown that 9 per cent cut-off rate is consistent with the Plan targets of 5.5 per cent growth of real GDP given the investment rate, ICOR and relative share of capital.[2] In fact, if we redo the exercise

2. $G = \frac{1/Y}{\Delta K/\Delta Y} = \frac{1/Y}{aK/Y} = \frac{r(1/Y)}{aR_K}$ where r is the cut-off ERR

Therefore, $r = \frac{r.R_k.G}{(1/Y)}$

Around 7th Plan, G = 5.5%, (I/Y) = 27.5%, a = 1.8 and R_k = 0.25% hence r = 9%

with current estimates of the parameters, the cut-off rate is likely to be around 8-9 per cent only.

Thus, it can be argued that if SSP has an ERR of more than 9 per cent it would lead to acceleration of national growth, which in itself is desirable and hence not considered separately. However, none of the economic appraisals of SSP has given this explanation explicitly before ignoring the growth effects of the project.

3.4 In considering the costs and benefits of a project from the whole society's angle, it is customary to ignore pecuniary benefits and costs representing transfers of existing resources from one social group to another. Such transactions do not add any net real value in the system. Usually only the real costs and benefits of the project are considered in SCBA. However, if distributional aspects are explicitly considered, inclusion of pecuniary costs and benefits can be justified. In that case, both costs and benefits have to be considered in nominal rather than real terms. This presents insurmountable problems of interpersonal utility comparisons and the whole concept of economic appraisal would run into serious problems. Generally, therefore, all costs and benefits are measured in real terms or at constant prices and pecuniary costs and benefits are not considered in SCBA (see Musgrave and Musgrave, 1989, Ch.9)[3].

Applying these four principles, various elements of costs and benefits of SSP are identified and listed below.

3.4A Costs: All costs related to SSP with a given design and details clearly specified are in real terms. Pecuniary costs are not considered as discussed above. These costs must be truly incremental and due to the project only. All the costs can be classified into categories like direct and indirect, and tangible and intangible.

3.4A[1] *Direct Tangible Costs:* All project costs are direct tangible costs. They include construction of the dam, dykes, canals, distribution system, drainage, canal-head power-house (CHPH), river bed power-house (RBPH), municipal and industrial (M&I) water use structures, *en-route* storage facilities etc. They would also include operations and maintenance (O&M) costs. Command area development costs

3. The critics of SSP have again faltered seriously on this count by arguing for inclusion of increased land prices leading to higher pecuniary cost of R&R (Paranjpye, 1990), which is effectively a transfer and not a real cost of SSP.

and catchment area treatment costs should also be considered direct tangible costs of the project as without them, the project benefits may not flow to the same extent.

Over and above these items, direct tangible costs of SSP would also include cost of the land inundated by the reservoir. Since such land would be permanently lost to society, its social cost should be considered. It includes both agricultural and forest land and products from such land, i.e. crops and forest produce.

All the appraisals of SSP done so far have considered almost all these direct tangible costs, except the construction costs of M&I water use structures. This was perhaps because they also did not consider the M&I water benefits seriously in their appraisal. Only SPIESR (1995) considered both costs and benefits of M&I water. Moreover, the critics of SSP and hence SPIESR insisted on incorporating within SSP cost the share of expenditure on ISP in MP that Gujarat was required to meet by the tribunal award. There are several objections to including this cost in the cost of SSP. First, ISP costs are neither direct nor indirect costs of SSP because ISP is a stand-alone independent project. SSP could exist even without ISP. Second, SSP is a multi-state project and not only a Gujarat project. Third, the tribunal award was only a judgment and did not represent meticulous estimation. Fourth, the overall impact of ISP on SSP should be judged by considering two scenarios: SSP with ISP and SSP without ISP, as is done by the World Bank in its appraisal. World Bank estimates show that the viability of SSP increases with ISP rather than without ISP.

3.4A[2] *Direct Intangible Costs:* These would include rare plant species, historical or archaeological monuments, village shrines and places of pilgrimage submerged in the reservoir, which would be permanently lost. This is an emotional cost. Very often such intangible costs are not evaluated and are excluded. What can, however, be done is to include the costs of shifting or relocating the monuments, shrines, temples etc. in the project costs. In the case of SSP these are so small and insignificant that all appraisals have practically ignored them. World Bank (1990) in its reappraisal attempted to include many of them. The physical inconvenience and trauma of families displaced by inundation are also direct intangible costs. These too were not properly evaluated and included in the economic appraisals of SSP.

3.4A[3] *Indirect Tangible Costs:* This would include the cost of abatement, particularly compensatory afforestation if the direct cost of inundated forest land is not considered. If it is considered, the cost of compensatory afforestation would amount to double counting. Another item to be included here is the cost of public health due to mosquitoes and viruses spreading from stagnant water in areas not having regular water without SSP. Costs of animal health should also be considered. Similarly, any waterlogging leading to salinity would also impose and indirect cost on society.

However, a major indirect cost of SSP is induced pressure on land due to resettlement of displaced persons from the area inundated by the reservoir. These are not direct costs because, except the land and the immovable capital assets on it, all other factors of production including labour and entrepreneurship are saved. They only have to be relocated and resettled. The cost of their transport and transactions for resettlement becomes an indirect cost of the project. Any amount of land given to these project affected people (PAPs) and the consequent increase in the price of the land transacted are not real net cost to the society. They are transfers and, if at all, represent pecuniary costs.

Another important item of indirect cost is the loss of fisheries downstream due to likely changes in the flow of the river with SSP. These are, if at all, tangible losses. Another type of indirect tangible costs of SSP on the society is the costs of certain businesses closing down or facing considerable reduction in demand. For example, supply of water by tankers would practically stop with SSP water becoming available. Similarly, the demand for diesel and coal would significantly decline with SSP because surface water would substitute ground water and hydro-power would substitute thermal power. Such indirect tangible costs are not adequately captured and estimated in the appraisals of SSP.

3.4A[4] *Indirect Intangible Costs:* SSP involves rehabilitation of the colonies and social groups living in the submerged area. Since most of this resettlement is compelled by the project, it inflicts psychological costs and pains on the people to be shifted and on those among whom they are rehabilitated. All appraisals of SSP recognised these intangible psychological costs of those who are resettled but none has identified similar costs to the societies accepting them. Moreover, the psychological costs of those rendered unemployed in the trades which would be adversely affected by SSP are also the indirect intangible costs of SSP. Another intangible cost is in terms of disturbance of ecological and carbon-oxygen balance through submergence of 37,533 ha. land in the reservoir.

3.4B Benefits: The benefits of SSP too must be identified as incremental real benefits to society as a whole, attributable solely to the project. They should be based on the differences between two scenarios—one with SSP and one without. Moreover, the social benefits should also include resources and real cost saved by society due to SSP. Like costs, benefits can also be classified in the same four categories.

3.4B[1] *Direct Tangible Benefits:* SSP's major outputs are the most direct benefits of the project. These are water and fisheries. Water has several alternative uses like (1) an input to agriculture, often referred to as irrigation benefits; (2) an input to animal husbandry; (3) industrial uses; (4) municipal and drinking water; (5) hydro-power generation; and (6) navigation and waterways. All these are highly productive and by and large they are mutually exclusive. The exact use pattern of water is generally decided on the basis of social priorities and intensities of needs. Socially optimal allocation based on the equal-margin principle would imply equal marginal returns to society in different uses of water.[4] Therefore, contrary to what some critics have argued, marginal deviations of the actual water use pattern from the pre-decided or 'desired' use pattern would not result in any significant change in the estimates of social benefits of the project. Critics of the project often overlook the fact that the main output of SSP is water and not specific uses of water. Therefore, if water is not put to a specific use, it is used for other purposes. Such a reallocation does not substantially reduce the overall

4. Let social Output = X and uses of water in Agriculture (A), Industry (I) and Power (P) add up to total available water. Then, the social planner must allocate water to the three uses so as to
Maximise $X = f(A, I, P)$
Subject to $W = A + I + P$
Applying Lagrangian Multiplier method, this is achieved when $\lambda = SMP_A = SMP_I = SMP_P$ where SMP is social marginal product of water in different uses.

benefit to society. Increased total output of water would lead to increased water availability for each of these uses, which represents tangible direct benefit of SSP. Increased fisheries also represent direct benefits. Similarly, savings on account of flood control in the downstream area due to the project also entail direct tangible benefits. Such savings are, however, not seriously estimated in the economic appraisals of SSP.

3.4B[2] *Direct Intangible Benefits:* Public parks on the dam site, sheer visuals of huge structures and grandeur and scenery of contrast in the water levels near the dam are some of the attractions for tourists. These are intangible benefits of the project. However, the biggest direct intangible benefit of SSP is the assured water supply for agriculture, animal husbandry, drinking and municipal uses and industrial uses. Reduction of uncertainties of water supply in a severely water-scarce area carries a tremendous intangible benefit to society. SSP would help considerably to achieve water security in the economy. Almost none of these intangible direct benefits of SSP are meticulously evaluated or included in any of the appraisals of the project. World Bank (1990) had, however, attempted to consider the favourable impact of SSP on environment and ecology, which represent direct intangible benefits.

3.4B[3] *Indirect Tangible Benefits:* The indirect tangible benefits of SSP include (1) increased flow of tourists; (2) increased level of ground water in the command area resulting in savings of energy as well as tangible public health benefits; (3) facilitation of several development projects inhibited and constrained by inadequate water availability; (4) increase in employment resulting from increased investment in the economy; (5) increased production of several commodities having forward and backward linkages with sectors like agriculture, animal husbandry, forestry, fishery, tourism, construction and power; and (6) increased fuel wood and fodder. Also, because of SSP, the quality of several crops would improve with soft surface water instead of hard contaminated ground water. Most of these indirect tangible benefits have either not been considered or have not been seriously evaluated in the economic appraisals of SSP so far. TECS and the World Bank have recognised some of them but have not attempted to estimate them rigorously. Moreover, the hydro-power generated by SSP is the cleanest source of energy entailing significant environmental benefits that are indirect and tangible. Another very important indirect tangible benefit, to Gujarat in particular, has been the experience of handling R&R through Land Purchase Committees. It has paved the way for land market reforms attracting huge investments to the state subsequently.

3.4B[4] *Indirect Intangible Benefits:* Reduced compulsion for seasonal and permanent migration of people and animals on account of increased availability of water is a major intangible indirect benefit of SSP. It has emotional, psychological and socio-cultural benefit for those who would have been compelled to migrate without SSP. Education of the children of such seasonal migrants would be a significant long term social benefit. Social stress resulting from artificial separation of family members having to migrate out and consequent social evils and crimes would be substantially reduced with SSP. Moreover, improvement in ground water quality and reduction in salinity ingress are important indirect intangible benefits. The major security concerns in a society are those of water, food, energy, and the threats due to poverty and destitution. Water security helps achieving the other three types of security. Since SSP would result in water security for at least the next 30 years, its positive effects on the other three are important indirect intangible benefits. SSP to a large extent is an indigenous designing and engineering feat. The experience gained by engineers, contractors, designers and workers represents enormous intangible benefits to society. It has helped build Brand India. It has to be recognised that problems of oustees or displaced persons, in terms of their psychological and social trauma in resettlement, would remain with or without the project. Persons and groups may differ but phenomenon of displacement and resettlement is common to all. From the distinguishing features of SSP discussed above, it can be safely inferred that the magnitude of such displacement with SSP is likely to be substantially lower than that without it.

4. Evaluation of Costs and Benefits of SSP

All the substantial economic appraisals of SSP—by TECS (1983), World Bank (1985, 1990, 1995), and SPIESR (1995) have followed standard methodologies widely used in the literature and practice to evaluate costs and benefits of such big projects. These methodologies comprise the use of economic prices or shadow prices to reflect 'true' scarcity value of resources and output. All these appraisals have used a social cost of capital at 12 per cent p.a. in real terms, rendering all distant future costs practically irrelevant. Such a high rate was justified on the ground that it was recommended by the Planning Commission and by experts. None of these appraisal exercises attempted to estimate the marginal Social Time Preference Rate (STPR) for the country, which would have been most relevant as the social discount rate. It is possible to estimate STPR for the Indian economy around mid-1980s from a management case (see Dholakia and Oza, 1999, Appendix 6-B). The estimate turns out to be only 5.4 per cent p.a. As against this, use of Social Rate of Return (SRR) of 12 per cent p.a. or cut-off rate of 9 per cent p.a. recommended by the Nitin Desai Committee of the Planning Commission appear too conservative and out of tune.

4.1 Project Costs

There are broadly two approaches to estimating the social costs and benefits of a big project—the UNIDO approach and the OECD approach (also known as Little-Mirrlees approach). All the three appraisals of SSP have preferred to use largely the OECD or L-M approach. Here all the costs and benefits of the project are systematically broken down into quantities of commodities. For instance, project costs are decomposed into cement, steel, petroleum products, explosives, power, mechanical equipment, electrical equipment, transport equipment etc., besides unskilled labour and various categories of skilled labour. Then, these commodity quantities are evaluated at their market prices for financial cost estimates and also at economic prices for the economic or social cost estimates. Economic prices would be obtained as the international (border) prices for tradable commodities after converting into rupees at the exchange rate that 'truly' reflects the value of the rupee in foreign currency. For non-tradable items of costs, the social cost of such resource commodities is estimated by adjusting their market prices for domestic distortions due to taxes/subsidies, monopoly, quantitative and price restrictions etc. Except tax/subsidy adjustment of the market prices, all other domestic distortions are corrected by applying the 'standard conversion factor' (SCF) to such adjusted market price to obtain the corresponding economic prices.

All the three economic appraisals of SSP have followed this route to estimate/evaluate the economic costs of SSP. They have taken an SCF of 0.8 for most of the small items and non-tradable commodities. Likewise adjustments were also made for estimating shadow wage rate to reflect the social cost of unskilled labour used in the project. The adjustment factor of 0.65 was applied to the actual labour cost to obtain the social opportunity cost of unskilled labour as was done in other projects by the World Bank.

4.2 Cost of Submerged Land

The social cost of submerged or inundated land was estimated as 'the economic value of output from land that will be forgone for ever. Hence, estimates of forest and agricultural output have been made, economic output obtained by using economic prices and the resulting magnitudes treated as costs to the project. Income expected during pre-submergence forest clearance was subtracted from project cost' (TECS, 1983: 18). Similar treatment was given to land under the canal. The other appraisals, by World Bank and SPIESR, also followed the same method. However, it should be noted that the correct estimation would be when economic value of the net output ascribable to the submerged land and immovable capital/property is taken instead of the gross value of output. Thus, the net output ascribable to the submerged land and immovable property on it would be only a fraction of the value added, comprising rent and a part of interest. But TECS took 12 per cent as the discount rate for arriving at the present value of the annuity. It is an unjustifiably high discount rate for the purpose and should be reduced to an STPR of 5-6 per cent. As a result, the net effect of lower value of output foregone and lower discount rate would not affect the cost estimates of submerged land significantly.

There is another count on which such an estimate of social cost could be highly overstated. While World Bank (1990) clearly recognised the benefits of the drawn-down agriculture on the long patch of 214 km along the length of the SSP reservoir, it did not make any effort to estimate them systematically. Of the other appraisals, TECS (1983) and SPIESR (1995) have not even recognised such benefits. Technically the social cost of submerged land should be net of the benefits of draw-down agriculture. Given the dimension of such benefits, it is quite possible that they might offset the gross social cost of submergence of land.

4.3 Catchment Area Treatment

Another element of social cost of SSP is associated with the treatment of the catchment area of the dam. TECS (1983) and World Bank (1985, 1990 and 1995) have ignored it on the ground that such treatments have *in-situ* benefits for the land and it was found elsewhere that such benefits are more than the cost of treatment. No doubt such treatment creates externalities for the dam, and so long as it is financially viable its exclusion can be justified. If the direct cost of land treatment is included on costs side, its direct benefits should be included on the benefits side. Since the latter exceed the former, both can be ignored in the interest of conservatism. However, SPIESR (1995) has not ignored the costs of catchment area treatment and has not included the *in-situ* benefits in the social benefits of the SSP. It has also wrongly included Gujarat's share of the costs of the IS project in MP in the social costs of SSP. Thus, SPIESR (1995) has grossly overstated the social costs of SSP.

4.4 Environment and Miscellaneous Items

TECS (1983) and World Bank (1985) had ignored the environmental costs and benefits of SSP, but the other appraisals have considered them explicitly. Again the World Bank (1990) considered the environmental aspects of SSP in the same principle of conservatism and SPIESR (1995) merely adopted those estimates. World Bank (1990) considered all small and missing elements in its earlier exercise along with the environmental aspects. Thus, it considered costs and benefits of the SSP under each of the heads of (1) wildlife; (2) parks and natural reserves; (3) compensatory afforestation; (4) minor forest produce like gums, dyes, liquors, medicines, and smoke leaves; (5) public health; (6) navigation; (7) monuments and village shrines; (8) rare plants; (9) carbon-oxygen balance through vegetation and biomass; (10) micro-climate—atmospheric moisture; (11) induced pressure on land; (12) flood control on the downstream 5,000 ha;

(13) salination; (14) ground water; (15) tourism; and (16) Loss of fish in the river and estuary.

Some loss of river and estuary fish is likely downstream, though not all loss of water volume is attributable only to SSP. The Bank report of 1990 has estimated the losses of *hilsa* and other fish on this account by year over the next 50 years. It also says that the silt loss in the estuary over time should be attributed to all the upstream dams.

Wildlife losses were estimated by linking them to forest losses (50%) after adjusting for the present trend of loss of habitat and increased real amenity value of wildlife over time. Wildlife benefits due to improved lakeshore environment were estimated again with increased amenity value over time. Similarly parks and nature reserves were considered for costs and benefits. Since no parks or reserves were getting inundated, the costs were treated as zero. Benefits to four parks were notionally and nominally considered due to increased water supply.

To avoid double counting, compensatory afforestation costs were not considered as the value of lost forest was already included in the submergence costs. Minor forest produce costs were relatively small due to degradation of forests in the submergence zone. They were considered at 20 per cent of the forest losses and were kept constant over time, although in all realistic scenarios, they would reduce over time.

Public health costs were estimated as cost of surveillance and control of water-related diseases, with operating costs at 50 per cent. The public health benefits were expected to be substantial owing to increased supply of reliable drinking water and washing water in the command area, and to increased food production coupled with income generation. However, the value of such benefits was conservatively (essentially arbitrarily) taken just to offset the costs after 20 years. World Bank (1990) felt that it could be a substantial underestimate of benefits of SSP.

Navigation costs in the estuary were highly uncertain and would occur, if at all, 35 years after the start of the project. Benefits from navigation, particularly in the long reservoir, were likely to be higher. Similarly, local transport benefits on roads associated with canals were ignored though such roads constituted a part of canal costs. The costs of shifting three larger temples were included as a contingency.

Rare plant losses were considered unlikely but were conservatively treated on par with wildlife.

The costs and benefits of carbon-oxygen balance and of micro-climate in terms of greater atmospheric moisture were estimated notionally by comparing the scenarios with and without SSP. It was argued rightly that benefits on these counts would far exceed the costs and were, therefore, ignored by World Bank (1990) in keeping with its conservative stance. The costs and benefits of changes in the induced pressure on land due to SSP were also estimated considering the situation with and without SSP. Again no hard estimates were made since benefits would far outweigh the costs.

Costs of flood control in terms of nutrient loss after floods were estimated per hectare for 5,000 ha. downstream. The benefits of flood control were arbitrarily taken as a lump sum amount per annum to reflect savings due to averted damage to farm equipment and averted loss of human and animal lives in floods.[5] The benefits were substantially underestimated. Similarly, salinity reduction due to increased fresh water percolation from the surface was practically not estimated. The costs of salination were set to zero because the project costs included both horizontal and vertical drainage to neutralise salinity costs. The ground water recharge benefits could be substantial in terms of saving on pumping costs in the command area. However, only 25,000 wells were assumed to be benefited by a marginal amount per well per annum. Similarly, tourism benefits could be substantial but were not estimated by the World Bank (1990). All these aggregated financial values were converted to economic values by applying a standard conversion factor of 0.8.

4.5 Irrigation Benefits

Net irrigation benefits were estimated in all the appraisals of SSP following a similar methodology. They differed only in details and models to predict exact numbers and in terms of base year prices. All of them considered two scenarios—with and without SSP. The

5. This amounts to putting a value on the human lives saved. However, World Bank (1990) has avoided assigning any such value explicitly. Without the project, human and animal lives would be routinely lost as floods were almost a regular occurrence.

difference in the net irrigation benefits between these was attributed to SSP. The four factors considered in order to calculate the net irrigation benefits were:

(a) Cropping pattern;
(b) Yields;
(c) Cost of cultivation; and
(d) Farm-gate prices by crop.

All these factors were considered for each of the 13 agro-climatic regions of the SSP command area. The cropping pattern and crop yields were projected based on past data of 15 years for the scenario without SSP. The projected yields were subjected to the constraint of minimum yield required to offset the cost of cultivation within the region. The yield rates and cropping pattern with SSP were estimated by considering the experience in an agro-climatically equivalent region with irrigation potential (SPIESR, 1995). Such estimates were also corrected with the help of expert opinion to avoid absurd outcomes. The farm-gate prices for different crops were considered after adjusting for inflation and taking a triennial average to avoid undue fluctuations. The corresponding economic prices were estimated using the standard L-M method. The cost of cultivation was taken in the financial terms and converted into economic terms again by the standard L-M method. The items considered for cultivation cost of each of the crops were: (a) human labour; (b) animal labour; (c) machinery; (d) seed; (e) manure; (f) fertiliser; (g) pesticide; (h) depreciation; and (i) interest.

The calculation procedures were based on a behavioural model and learning curve hypothesis of 3 years for adoption and full yield response with SSP. The net irrigation benefits were thus generated for each year over 50 years time.

4.6 Fuelwood and Fishery Benefits

These were ignored by the TECS (1983) and World Bank (1985) appraisals. However, World Bank (1990 and 1995) considered them explicitly and SPIESR (1995) adopted the same estimates, making changes for the base year prices. The World Bank (1990) estimated the incremental growth of trees along the canal side and on farm with SSP, compared to the scenario without SSP. Financial values (price) per tonne of fuelwood and per piece of pole were considered and costs were deduced. The net benefit from firewood was then converted to economic prices by applying the SCF of 0.8.

The fishery benefits were considered in terms of (a) the reservoir itself; (b) canals; (c) development of fish-ponds; (d) development of prawn ponds; (e) development of shrimp production using drainage water to regulate the salinity in brackish water ponds near the sea; and (f) cage fisheries in both reservoirs and canals (World Bank, 1990). Yields were assumed to stabilise over time. Costs of development for all of these were estimated and deduced. The financial values were converted into corresponding economic net benefits by using an SCF of 0.8 (SPIESR, 1995).

4.7 Power Benefits

The approach and methodology followed for estimating the power benefits of SSP differ substantially. TECS (1983) and SPIESR (1995) followed the supply approach, i.e. the opportunity cost approach. Thus, they estimated the power benefits in terms of costs averted in supplying the same amount of power by the most practical alternative technology. It may be noted, however, that the most practical alternatives, of thermal plant for power supply on firm basis and gas plant for peaking purpose, are not necessarily the best and socially cheapest way to supply the same power in the absence of SSP. On the other hand, World Bank (1985, 1990 and 1995) used the demand approach, i.e. willingness-to-pay approach. TECS (1983) too claimed that they had estimated the power benefits through the demand approach. However, they did not provide the estimate and used only the opportunity cost approach to evaluate the power benefits. Substantial power benefits are accrued from Riverbed Power House (RBPH) and Canal Head Power House (CHPH) with installed capacity of 1200 MW and 250 MW respectively.

The alternatives considered by TECS (1983) and SPIESR (1995) differed in that TECS considered a 210 MW thermal plant for power supply on firm basis and 150 MW gas plant for peaking purpose, but SPIESR considered a 615 MW gas plant for peaking purposes. Other than this the details of their estimation of power benefits were the same. Neither of these two appraisals considered the social cost of pollution due to the alternative method of supplying the power. It is well known that hydro-power entails significantly lower environmental costs than its alternatives. This major limitation of the supply approach was not recognised by TECS (1983) and SPIESR (1995). Moreover, inasmuch as power is not traded across borders and is an item of benefit rather than cost of the project, the L-M methodology suggests that it should be evaluated through a willingness-to-pay approach and not through opportunity cost approach. The World Bank (1985, 1990 and 1995) had considered the right approach.

World Bank (1985) had estimated the willingness to pay for power by considering the consumer surplus in each consumer category. In order to arrive at this, the additional economic cost of alternative of private supply for each consumer category was estimated. Half of this additional economic cost was then added to economic value of the average tariff in each consumer category to derive the power benefits. This was based on the assumption that price elasticity of demand for power was unity. In practice, price elasticity at low levels of electricity tariffs in less developed countries is well below unity. Thus, the estimated consumer surplus by World Bank (1985) underestimated the true surplus and hence the value of power benefits from SSP.[6] The alternatives considered were electric generation sets for private/industrial users and irrigation using diesel pumps. Moreover, the power benefits of SSP critically depend on the upstream developments, particularly the Indira Sagar Project in MP.

---

6. Elasticity of demand (e) in terms of power, quantity (Q) and its price (P) is—

$$e = \frac{\Delta Q}{\Delta P} \frac{(P_1 + P_2)/2}{(Q_1 + Q_2)/2} < 1$$

Therefore, $\Delta Q\,(P_1 + P_2)/2 < \Delta P(Q_1 + Q_2)/2 = \Delta P\left(Q + \frac{\Delta Q}{2}\right)$

$$= Q_1 \Delta P + \frac{\Delta P \Delta Q}{2}$$

4.8 Municipal and Industrial (M&I) Water Benefits

TECs (1983) did not consider these benefits and associated incremental costs. World Bank (1985) did, but the valuation was very arbitrary and unrealistically low. When it was revised at 1989 prices, it was evaluated at the rate of Rs. 0.74 per cubic metre at the start of the project, rising to Rs. 3.68 per cubic metre over 30 years which would come to less than 3 paise per gallon. World Bank (1990) recognised that it was a substantial underestimate of the willingness to pay in view of 'the devastating impact on these areas of recent droughts which required trucking of water to villages, 20 minutes village water supply every 3 days in many villages and resulted in water riots in some areas as well as very high losses of livestock.' However, it did not make any effort to revise the estimates.

SPIESR (1995) differed substantially from previous appraisals on this count. They attempted to estimate M&I water benefits and also the costs of structures these needed. The approach followed, however, was supply based and not demand based. They estimated the least cost alternative to supply the same amount of water in a scenario without SSP. With SSP, this cost would be averted, which represented the benefit of SSP to society. Demand for water was estimated based on population projection and likely changes in economic structure. For Saurashtra and Kachchh, the cost of supplying Narmada water was compared with the cost of supplying an equivalent quantity through desalination, which was the least cost alternative.

4.9 Indirect Benefits

TECS (1983) made the only effort to estimate some indirect benefits of SSP in terms of employment generation. However, neither their appraisal nor those of the others addressed these in terms of increased output, income, consumption, savings etc. TECS estimated employment generation due to SSP considering direct employment during the implementation or construction phase and incremental agricultural output. It did not consider increased output of power and the resulting increase in industrial and commercial activities. The estimates of additional employment were made using the national input-output table prepared by the Planning Commission for the year

1979-80, for 89 × 89 commodity sectors. It was aggregated into 46 × 46 commodity sectors. The Leontief inverse was taken along with employment coefficients by sector to derive estimates of indirect employment generation on account of SSP.

5. Economic Rate of Return (ERR) for SSP

From the discussion above, it is clear that none of the economic appraisals or the SCBAs of SSP conducted so far is (1) comprehensive in coverage; (2) meticulous in detail; or (3) technically and theoretically perfect. Each suffers from one limitation or another. None of them have attempted serious estimation of (a) some of the direct intangible costs and benefits; (b) several indirect tangible benefits; and (c) most of the indirect intangible costs and benefits. However, all have followed certain basic principles of SCBA to impart credibility and reliability to their estimates. They have carried out the SCBA with different base year prices—1981-82 (TECS, 1983), 1984 (World Bank, 1985), 1989 (World Bank, 1990) and 1991-92 (World Bank, 1995 and SPIESR, 1995). The base year changes in the prices are not expected to make any major difference in the results and these different SCBA exercises have confirmed it.

However, the exercises not only considered changes in the base year prices but also took into account whatever minor or major changes in the design and time-pattern of costs and benefits were made over time. A common feature of all these SCBA exercises was that each one had taken the most conservative approach to estimate the benefits of SSP and considered the costs most liberally. In doing so, they did not care if some theoretically valid and correct treatment was not given to items. As a result, the final conclusion or the estimate of ERR and benefit-cost ratio for the project emerging from their studies were definitely on the low side. The 'true' ratios would certainly be much higher than what they had estimated.

The first formal economic appraisal (TECS, 1983) estimated the benefit-cost ratio on discounted present value basis at 1.39 at market prices and 1.84 at economic prices. The estimate for the ERR (at economic prices) was 18.3 per cent at constant prices. Adding the inflation rate of about 8 per cent p.a. would yield a nominal comparable rate of return of about 26.3 per cent.

The World Bank (1985) did not consider benefits and costs of SSP in terms of environment, fuelwood and fisheries. These were added in the revised exercise (World Bank, 1990). The original SCBA by World Bank (1985) estimated the ERR to be 13 per cent (at economic prices). With revisions of the estimates incorporating additional items of costs and benefits as well as the latest information on time-pattern and design, World Bank (1990) estimated the ERR to be 12 per cent—well above the acceptable (cut-off) rate for a large multipurpose regional project with substantial multiplier effects on the economy. World Bank (1990) recognised that the Nitin Desai Committee of the Planning Commission had recommended 9 per cent as acceptable rate with 7 per cent for drought prone areas.

World Bank (1990) also estimated the ERR for SSP at 12.5 per cent taking costs already incurred by then as sunk costs. Similarly, it found ERR of about 16 per cent if SSP was considered only as a power project. Since there was considerable delay in implementing ISP (at that time it was called Narmada Sagar Project or NSP) in MP, World Bank (1990) also estimated a scenario for SSP alone. The ERR for SSP with NSP was 12 per cent and SSP without (or delayed) NSP was 11 per cent. Thus, NSP would increase benefits from SSP rather than reducing them as some of the critics of SSP argued. However, subsequent revision of the estimates at 1991-92 prices by the World Bank (1995) gave 12 per cent ERR for SSP assuming a 5-year delay in NSP. If SSP was considered only as a power project, the ERR would be 14 per cent; and if all costs till that date were taken as sunk costs, it would be 17 per cent.

Another extremely important point made by the World Bank (1990) was about implementation delays and their impact on ERR. If the whole project got delayed by two years, there would be no significant effect on ERR because all costs and benefits would get delayed by two years. But any delays in the final-stage development of the command area distributaries and on-farm elements could lead to a significant adverse impact. However, the 'multi-purpose' character of SSP could make its economics quite stable even with such delays: if irrigation got delayed, more water would be available for power generation, which was a higher value proposition (World Bank, 1990).

Finally, SPIESR (1995) considering M&I water costs and benefits more realistically compared to the earlier appraisals and also incorporating some cost elements, (which should ideally not have been included), estimated the benefit-cost ratio as well as ERR at financial prices and at economic prices considering eighteen different scenarios with three alternative project costs, two alternative M&I water benefits and three alternative irrigation efficiencies for ERR. For B-C ratio, they additionally considered two alternative social discount rates of 10 per cent and 12 per cent, taking the number of alternatives to 36.

Three different project costs were used corresponding to different assumptions of technology used in the canal system, with the lowest project cost being comparable to the one used by World Bank (1990) and TECS (1983). Similarly, three different irrigation efficiencies—45 per cent, 53 per cent and 61 per cent were used. The M&I water benefits were calculated with two alternative assumptions without SSP: (1) only two-third of the additional regional demand of water would be met by desalination; and (2) full demand would be met by desalination.

The ERR (at economic prices) varied from 16.77 per cent to 21.88 per cent with about 19 per cent as the mean scenario value in the exercise by SPIESR (1995). The IRR at financial prices varied from 14.69 per cent to 20 per cent with about 16 per cent as the mean scenario value. The B-C Ratio varied from 1.31 to 2.79 at financial prices with about 1.7 as the mean scenario value. However, at economic prices the B-C ratio varied from 1.59 to 3.29 with about 2.3 as the mean scenario value.

It is clear from the above review of the economic appraisals of SSP, carried out by different independent professional bodies, that with stringent treatment of benefits and very liberal consideration of all possible costs, the project turns out to be socially and economically highly viable. The most conservative estimates of ERR also point to a distinct possibility that SSP would lead to acceleration of economic growth in the state and the country. This already seems to be happening (see Dholakia, 2010: 1-14).

Finally, the most contentious issue, of rehabilitation and resettlement of oustees of SSP in particular and of any large developmental project in general, needs to be viewed from the perspective of social cost benefit analysis. Generally it is extremely difficult (if not impossible) to find developmental projects where everybody in society gains and nobody loses. In most cases, such projects result in many people gaining, some people not affected directly or indirectly, but a few people definitely losing. This happens because society by and large attains equilibrium in Pareto optimal conditions. In such a state, nobody can be made better off without somebody becoming worse off. Therefore, the distribution of benefits and costs of the project becomes critical and every developmental project is most likely to have specific distributional implications. SSP is no exception. If there is no unanimously agreed mechanism to resolve this matter, interpersonal utility comparisons will make conceptually meaningless the whole exercise of estimating costs and benefits of a project to the society. The only alternative to extricate the exercise and arrive at a meaningful solution is to promote a social bargain between the potential losers and potential gainers from the project, based on the Kaldor-Hicks Compensation Principle. According to this principle, the gainers should compensate (win over) the losers through mutually acceptable voluntary bargain *before the project starts*. The R&R package offered in Gujarat to the project affected people (PAP) in SSP can be seen in this context. However, the issue became contentious got entangled in legal controversies because the PAP in MP and Maharashtra did not receive similar benefits before the start of the project. Thus, SSP presents a good illustration of what happens when the principles of sound SCBA are not followed in practice. It may only result in avoidable controversies, unnecessary opposition and huge waste of invaluable time.

References

Acharya, S., S.P. Punalekar, N.K. Gourah and N.V.V. Char, (1992): *Report of Fact-finding Team Appointed by Ministry of Water Resources on Manibeli*, Government of India. New Delhi.

Ahluwalia, M.S. (2000): "Economic Performance of States in Post-Reforms Period," *Economic and Political Weekly*, May 6-12.

Ahmedabad Municipal Corporation (AMC) (2010): *Information on Reforms Achieved for Consideration under The Awards for Excellence In Municipal Reforms and Service Delivery Improvements*, Report Submitted to Ministry of Urban Development (MoUD), Government of India.

Alagh, Y.K. (2010): "A Sardar Sarovar riddle", *The Financial Express*, Ahmedabad Edition, April 19.

Alagh, Y.K., Mahesh Pathak, and D.T. Buch (1995): *Narmada and Environment—An Assessment.* Har Nand Publications. Delhi.

Alagh, Y.K., R.D. Desai, G.S. Guha, and S.P. Kashyap (1995): *Economic Dimensions of the Sardar Sarovar Project.* Har Anand Publications. SPIESR. Ahmedabad.

Alexander, K.C., R.R. Prasad and M.P. Jahagirdar (1991): *Tribals, Rehabilitation and Development,* Rawat Publications. Jaipur.

Allen, R.G., L.S. Willardson and H. Frederiksen (1998): "Water Use Definitions and Their Use for Assessing the Impacts of Water Conservation" in J.M. de Jager; L.P.Vermes and R. Rageb (eds.), Proceedings ICID Workshop on *Sustainable Irrigation in Areas of Water Scarcity and Drought*, Oxford, England. September, 11-12.

Alvares, C. and R. Billorey (1988): *Damming the Narmada*. Natraj Publishers. Dehra Dun.

Alvares, Claude and Ramesh Billorey (1987): "Damming the Narmada: The Politics Behind the Destruction," *The Ecologist*, Vol. 17, No. 2.

—(1988): *Damming the Narmada: India's Greatest Planned Environmental Disaster,* Natraj Publishers. Dehradun.

—(1988): *Damming the Narmada,* Third World Network. Penang, Malaysia.

Amarasinghe, Upali *et al.* (2005): *Spatial Variation in Water Supply and Demand across River Basins of India,* Research Report; 83, International Water Management Institute, Colombo.

ARCH-Vahini (1988a): *Displacement in Sardar Sarovar (Narmada) Project: A Gujarat Experience*, Mangrol, Rajapipla. April.

— (1988b): *Sardar Sarovar Oustees – Which Way to Go: An Activist's Dilemma*, Mangrol, Rajpipla. September.

— (1991): "Resettlement and Rehabilitation in SSP: Are the Critics Right?" (*Mimeo*).

— (1992a): *Is R&R Impossible in SSP: The Role of NBA (India), EDF (USA) and MOEF (India): A Submission to Morse Mission.* Mangrol, Rajpipla. February.

— (1992b): *A Critique of Memorandum Submitted to the World Bank by the EDF and Other NGOs* on *March 20, 1991,* Mangrol, Rajpipla. September.

— (1992c): *Letter to Mr. Lewis Preston, President, World Bank on Morse Report,* Mangrol, Rajapipla.

— (1992d): *Letter to David Jones, Associate Director, Oxfam U.K. on Morse Report*, Mangrol, Rajpipla. September.

— (1992e): *The Independent Review: A Search for Truth?* Mangrol, Rajpipla. November.

— (1993a): *Morse-Berger Review of Sardar Sarovar Project: A Counterview*, Mangrol, Rajpipla. January.

— (1993b): *The SSP Withdrawn from the World Bank: ARCH-Vahini's Views*, Mangrol, Rajpipla. April.

— (1993c): *The Sardar Sarovar Project: An Intellectual Fashion*, Mangrol, Rajpipla. July.

— (1993d): *Human Rights in the Sardar Sarovar Project: A Submission to Fact Finding Committee of PUCL*, Mangrol, Rajpipla. October.

— (1994): *A Note of PAPs in M.P. who are Entitled for Agricultural Land for Research,* Mangrol, Rajpipla. March.

— (1995): *R&R in SSP: Are the Critics Right?* in Pathak, Mahesh (ed.) *Development Imperatives and Environmental Concerns.* Agro Economic Research Centre. Vallabh Vidyanagar.

ASCE (1987): *Planning, Operation, Rehabilitation and Automation of Irrigation Water Delivery System,* Proceeding of ASCE Symposium, Portland, Oregon.

— (1993): "Special Issue of Canal System Modelling", *Journal of Irrigation and Drainage,* No. 119 (4).

Barot, J.M. (1996): *Occurrence of Salinity in Gujurat in Reaching the Unreached: Challenges for the 21st Century*, 22nd WEDC Conference, New Delhi, India.

Baviskar, Amita (1991): "The Researcher as a Pilgrim," *Lokayan Bulletin*, Volume 9 (3-4). March.

Berg, Elliot and Mary M. Shirley (1987): *Divestiture in Developing Countries,* World Bank Discussion Paper No. 11, Washington, DC.

Bhaskaracharya Institute for Space Applications and Geo-informatics (2009): *Geographical Information System for Sardar Sarovar Project (SSP) Command Area Phase I: Project Proposal*, Gandhinagar.

Bhatia, Bela (1992): "Lush Fields and Parched Throats – Political Economy of Groundwater in Gujarat", *Economic and Political Weekly*, December 19-26.

— (1997): "Forced Eviction in Narmada Valley" in Jean Dreze; M. Samson and S. Singh (eds.) *The Dam and the Nation,* Oxford University Press. Delhi.

Border Environment Cooperation Commission (2003): "Hidalgo County Irrigation District

Brewer, J., S. Kolavalli; A. H. Kalro; G. Naik; S. Ramnarayan; K.V. Raju; and R. Sakthivadivel (1999): *Irrigation Management Transfer in India: Policies, Processes, and Performance,* A collaborative study by the Indian Institute of Management, Ahmedabad, India and the International Irrigation Management Institute, Colombo, Sri Lanka. Oxford & IBH Publishing Co. Pvt. Ltd, New Delhi.

Bribiesca, Sanchez (1967): *9th International Commission on Large Dams.* Turkey.

Bruns, Bryan Randolph and Ruth Meinzen-Dick (2000): *Negotiating Water Rights,* Sage, New Delhi.

Bureau of Indian Standards (1986): *Criteria for Earthquake-Resistant Design of Structures* in *Indian Standard IS 1893: 1984.* New Delhi.

CAG Report (2000 and 2001): Comptroller and Auditor General Report 2002, Government of India.

Census of India (2001): *Housing and Amenities Table, Gujarat.*

Central Board of Irrigation and Power (1993): *Canal Automation*, Publication No. 238, New Delhi.

Central inland Capture Fisheries Research Institute (1993): *Studies on Fish Conservation in Narmada Sagar, Sardar Sarovar and it's Downstream*, Report submitted to Narmada Control Authority (NCA), Indore, Government of India.

Central Soil and Water Conservation Research and Training Institute (n.d.): Interim Report on *Rainfall Runoff and Sediment Yield Measurement in Catchment Area of Sardar Sarovar Project lying within Gujarat.* Vasad.

CEPT - CRISIL (2006): *Public Private Partnership: A Study of Potential Benefits in the Sardar Sarovar Narmada Project* (Unpublished), Submitted to Sardar Sarovar Narmada Nigam Limited, Gandhinagar.

Cernea, Michael (1995): *Keynote Address: Understanding and Preventing Impoverishment from Displacement* in *International Conference on Development Induced Impoverishment.* University of Oxford. Oxford.

— (1991): "Involving Settlement: Social Research, Policy and Planning" in Michael M. Cernea (ed.) *Putting People First: Sociological Variables in Rural Development*, Oxford University Press, Second Edition. New York.

— (1996): "Public Policy Responses to Development-induced Population Displacements," *Economic and Political Weekly*, Vol. 31 (24).

Chakravorty, Ujjayant and Chieko Umetsu (2003): "Basinwide Water Management: A Spatial Model", *Journal of Environmental Economics and Management*, 45 (2003): 1-23.

Chambers, Robert (1988): *Managing Canal Irrigation: Practical Analysis from South Asia,* Oxford and IBH, New Delhi, India.

Chandra, A. and C. Sudhir (2010): *A Study of Kolhapur Lift Irrigation Co-operatives*, MTS Report, Institute of Rural Management, Anand.

Choudhury, N. and V. Kher (2006): "Public-Private Partnership in Surface Water Irrigation: A Case of Kolhapur", IWMI Tata Water Policy Program, Anand (unpublished research).

CICFRI (n.d.): *Studies on fish Conservation in Narmada Sagar, Sardar Sarovar and its Downstream.* Barackpore.

CMIE (2010): *India Harvest Database*, Centre for Monitoring Indian Economy.

Commissioner of Fisheries, Govt. of Maharashtra's letter No. Fish/T/011403/2168 to Secretary, Narmada Ghati Development Department, Government of MP.

Commissioner of Health, Medical Services and Education (1993): *Environmental Impact Assessment Studies on water Related Diseases in SSP command Area Including the Area Downstream of the Narmada Dam.* Government of Gujarat. Gandhinagar.

Cook, Paul and Colin Kirkpatrick (eds.) (1988): *Privatization in Less Developed Countries,* St.Martin's Press, New York.

Cooter, Robert D. (1997): "The Rule of State Law and the Rule-of-law State: The Economic Analysis of the Legal Foundations of Development" in Micheal Bruno and Boris Pleskovic (eds.) *Annual World Bank Conference on Development Economics 1996*, World Bank, Washington, D.C.

Cowan, L. Gray (1990): *Privatization in the Developing World,* Greenwood Press, New York.

Cullet, Philippe (2007): *The Sardar Sarovar Dam Project: Selected Documents*, Ashgate Publishing Company. Hampshire. U.K.

Dalua, A.K. (1993): *Environmental Impact of Large Reservoir Projects on Human Settlement*, Ashish Publishing House. New Delhi.

Dam Safety Panel (1997): *Report of the Thirty Fourth Meeting of Dam Safety Panel, Sardar Sarovar (Narmada) Project,.* Sardar Sarovar Narmada Nigam Limited. *Gujarat.*

DANIDA (1996): *Community Participation in Rural Drinking Water and Sanitation Projects: A Comparative Study of Five Projects in India*, RWS & Sanitation Group, SA, UNDP-WB & IIMA.

Das, Biswaroop (1983): "Studies on Rehabilitation of Submerging Villages – Vadgam" *(Mimeo)*. Centre for Social Studies. Surat.

— (1986): "Submerging Tribal Belt in Gujarat: Spatial Variation and Stratification," *State and Society*. 6(1).

— and A.S. Charan (1983): "Cost of Submergence," *(Mimeo)*. Centre for Social Studies. Surat.

Datta, Samar K. (ed) (2003): *Cooperatives in Agriculture – A Millennium Study* (Volume 24), Academic Foundation, Delhi.

— *et al* (forthcoming): *Developing India's Strategies Response to the Global Debate over Fisheries Subsidies*, Chapter 4, Allied Publishers, Delhi.

Daud Committe (2001): *Report of the Committee to Assist the Resettlement and Rehabilitation of the Sardar Sarovar Project-Affected Persons*, Government of Maharashtra.

Dave Kapil (2010): "Half of Gujarat will live in cities". *Daily News and Analysis*, June, 2.

Davies, Susanna and Naomi Hossain (1997): "*Livelihood Adaptation, Public Action and Civil Society,*" *Working Paper 57*, IDS, Brighton.

De Fraiture, Charlotte and Chris Perry (2002): "Why is irrigation water demand inelastic at low price ranges?" Paper presented at the conference on *Irrigation Water Policies: Micro and Macro Considerations*, Agadir, Morocco, June, 15-17.

Department of Botany (1983): *The Sardar Sarovar Narmada Project Studies on Ecology and Environment.* M.S. University. Baroda.

— (1992): *Eco Environmental and Wildlife Management Studies on SSP Submergence Area in Gujarat.* M.S. University. Baroda.

Department of Earthquake Engineering (1982): *Design Earthquake Parameters for Navagam Dam Site in Gujarat.* University of Roorkee.

Department of Environment and Forests (1986): *Environmental Aspects of Narmada Sagar and Sardar Sarovar Multi-purpose Project*, URL: http://www.ielrc.org/content/c8601.pdf. December.

Desai Vaidya, Prerana (2010): "Impact Of Sardar Sarovar Project: A Field Evaluation." *Seminar on Evolution and Growth of Sardar Sarovar, the Lifeline of Gujarat,* CEPT University, Ahmedabad.

Desai, G.M., Gurdev Singh and D.C. Sah (1979): "Impact of Scarcity on Farm Economy and Significance of Relief Operations," *(Mimeo)*. Indian Institute of Management. Ahmedabad.

Desai, N.B. (1990): "*Influence of Sedimentation on Service Life of Sardar Sarovar,*" *Indian Journal and Power and River Valley Development, Special Number on Sardar Sarovar Narmada Project,* September (pp. 249-253)

Desai, S.J. and M.B. Joshi (2008): *Narmada Water Plays Its Role – Capturing Initial Trends from Gujarat,* Presentation, Sardar Sarovar Narmada Nigam Ltd., Gandhinagar.

Deshpande, N. (2009): *Displaced by the Dam: Living on the Edge,* URL (http://www.indiatogether.org/2009/oct/hrt-sardar.htm), October.

Dharmadhikary, S. (2005): *Unravelling Bhakhra: Assessing the Temple of Resurgent India*, Manthan Adhyayan Kendra, Bhopal.

Dhawan, B.D. (ed.) (1990): *Big Dams: Claims, Counter Claims,* Commonwealth Publishers, New Delhi.

Dholakia, Ravindra H. (1985): *Regional Disparity in Economic Growth in India*. Himalaya Publishing House. Bombay.

—(1994): *Spatial Dimensions of Acceleration in Economic Growth in India* in Economic and Political Weekly. Volume 29, Issue 35 (pp. 2303-09)

—(2010): *Has Agriculture in Gujarat Shifted to High Growth Path?* in Dholakia, Ravindra H. and Samar K. Datta (eds.): *High Growth Trajectory and Structural Changes in Gujarat Agriculture*, Macmillan Publishers.

—and Samar K. Datta (eds.) (2010): *High Growth Trajectory and Structural Changes in Gujarat Agriculture,* Macmillan Publishers.

—and Ajay N. Oza (1999): *Microeconomics for Management Students*, 2nd Edition. Oxford University Press. Delhi.

Dixon J.A., L.M. Talbot and G.J.M. Le Moigne (1989): *Dams and the Environment: Considerations in World Bank Projects*, World Bank Technical Paper Number 110, Washington D.C.

Domanski, Tomasz (1992): "Development of Small Private Companies and Their Marketing Activities," in *Journal of Business Research,* Vol. 24, pp. 57-65.

Falkenmark, Malin and Carl Widstrand (1992): "Population and Water Resources: A Delicate Balance." *Population Bulletin,* Population Reference Bureau: Washington, DC.

FAO (2008): *Report of the Lake Victoria Fisheries Organization and FAO - National Stakeholders' Workshops on Fishing Effort and Capacity on Lake Victoria (2006),* FAO Fisheries and Aquaculture Report No. 817, Food and Agriculture Organization of the United Nations, Rome.

Fernandez, A.P. (1990): "No Dam is an Island" in A P Fernandez (ed.) *Workshop on Rehabilitation of Persons Displaced by Development Projects: A Report,* Institute for Social and Economic Change and MYRADA. Bangalore.

Final Order and Decision of the Narmada Water Dispute Tribunal (NWDT) Gazetted on December 12, 1979, Source: nca.gov.in/forms_pdf/nwdt_finalorder.pdf

Fisher, William F. (2001): "Diverting the Water: Revisiting the Sardar Sarovar Project", *International Journal of Water Resources Development*, 17 (3): 303-314.

Fitch, Lyle C. (1988): "The Rocky Road to Privatization," in *American Journal of Economics and Sociology*, Vol.47, No. 1, pp. 1-14.

Frederick, K D (1992): *Balancing Water Demand with Supplies: The Role of Management in a World of Increasing Scarcity*, Technical Paper 189, Washington DC: World Bank.

GoI (1987): *National Water Policy: 1987*, Government of India, Ministry of Water Resources. New Delhi.

GoI (2002): *National Water Policy: 2002*, Government of India, Ministry of Water Resources, New Delhi.

Goldsmith, Edward, and Nicholas Hildyard (1984): *The Social and Environmental Effects of Large Dams*, Sierra Club Books. San Francisco.

GOPA Consultants (1991): *Narmada River Basin Development Project* – Report submitted to SSSNL, Government of Gujarat. http://www.nvda.nic.in/env.htm

Government of Alberta (2003): "Alberta's irrigation infrastructure," http://www1.agric.gov.ab.ca/$department/deptdocs.nsf/all/irr7197/$FILE/irrigationinalta-part2.pdf (accessed on 11th November, 2009).

Government of Gujarat (1987): *Compilation of Government Resolutions for Rehabilitation and Resettlement Policy for Oustees of Sardar Sarovar Project*, Narmada Development Department. Gandhinagar.

— (1992): *No Cancelling Tomorrow: A Critique of the Report of the Independence Mission of Gujarat*, Gandhinagar.

— (1992): *Comment on the Report of the Independent Review Mission on SSP.* Gandhinagar.

— (1997): *Water Supply & Sanitation, 9th Five-Year Plan (1997-2002)*, Narmada Water Resource & Water Supply Department, Government of Gujarat, Gandhinagar.

— (1998): *Infrastructure 2010: Master Plan on Drinking Water Supply*, Narmada Water Resources and Water Supply Department, Gandhinagar.

— (2000): *Drinking Water Scarcity 2000-2001*, Narmada Water Resources & Water Supply Department, Gandhinagar.

— (2002): *10th Five-Year Plan (2002-2007): Water Supply & Sanitation*, Narmada Water Resource & Water Supply Department, Government of Gujarat, Gandhinagar.

— (2007): *Gujarat Water Users' Participatory Irrigation Management Act*, Government of Gujarat, Gandhinagar.

— (2009): *First Approximation report to increase groundwater resources through artificial recharge in Gujarat State*, Narmada Water Resources, Water Supply and Kalpsar Department, Gandhinagar.

— (2010): *Report of the Expert Group for Strategy for the Accelerated Development of Sardar Sarovar Project Command Area*, Sardar Sarovar Narmada Nigam Ltd., Gandhinagar (unpublished report).

Government of India (1978): *The Report of the Narmada Water Dispute Tribunal with its Decision*, Vols. I and II. New Delhi.

— (1990): *Approach to Eighth Five Year Plan*, Planning Commission. New Delhi.

— (1992a): *Eighth Five Year Plan (1992-97), Vol. II: Sectoral Programmes of Development*, Planning Commission. New Delhi.

— (1992b): *Report of the Five Member Group Set up by the Ministry of Water Resources to Discuss Various Issues Relating to the Sardar Sarovar Project*, Vols. I and II. New Delhi.

— (1987): *National Water Policy, 1987*, Government of India, Ministry of Water Resources.

— (2002): *National Water Policy, 2002, Government of India, Ministry of Water Resources.*

— (1995): *Report of the Committee on Private Sector Participation in Major and Medium Irrigation Projects*, Ministry of Water Resources, New Delhi, India.

— (2002): *National Water Policy,* Ministry of Water Resources, New Delhi, India.

— (2005): *Report on 3rd Census of Minor Irrigation Schemes*, Ministry of Water Resources, Minor Irrigation Division, New Delhi.

Government of Madhya Pradesh (1992): *Action Plan of R&R of Oustees of MP (SSP), Parts I and II,* Narmada Valley Development Authority.

Guggenheim, Scott (1990): "Development and Dynamics of Displacement" in A.P. Fernandez (ed.) *Workshop on Rehabilitation of Persons Displaced by Development Projects: A Report*, Institute for Social and Economic Change and MYRADA. Bangalore.

Gujarat Infrastructure Development Board (2009): *BIG 2020 – Blueprint for Infrastructure in Gujarat*, Gandhinagar.

Gujarat PUCL (1992): Human Rights Issues at Manibeli: Fictions and Facts *(Mimeo),* Report by Gujarat PUCL on behalf of Human Rights Group of Narmada Abhiyan.

— (1993), "Drowning of Manibeli," *PUCL Bulletin*, Vol. 12(7), Report by M. Mohanty, C.C. Kalara on behalf of Human Rights Campaign on Manibeli.

Gulati, Ashok; Tushaar Shah and Ganga Shreedhar (2009): *Agriculture Performance in Gujarat since 2000: Can it be a* Divadandi *(Lighthouse) for other States?* International Water Management Institute, Anand; and International Food Policy Research Institute, New Delhi.

Gupta, Rajiv K (2001): "Human Rights Dimension of Regional Water Transfer: Experience of the Sardar Sarovar Project," *International Journal of Water Resources Development*, Vol. 17 (1).

GWRDC (n.d.): *Ground Water Resource Estimation.* Retrieved from http://www.gwrdc.gujarat.gov.in/Gwremaps/GWRE_B.K..htm

GWSSB (1983): *Water Demand for Non-agriculture Use from Narmada Project.* Gandhinagar.

GWSSB (1985-2005): *Master Plans*, Gujarat Water Supply and Sewerage Board, Gandhinagar.

— (2003): *Sardar Sarovar Canal Based Drinking Water Supply Master Plan*, Gujarat Water Supply & Sewerage Board, Gandhinagar.

Hague Deceleration (2000): "Ministerial Declaration of The Hague on Water Security in the 21st Century," *Second World Water Forum*, Hague. March.

Hanke, S.H. (ed.) (1987): *Privatization and Development*, Institute for Contemporary Studies Press, San Francisco.

Haskoning (1999): *Netherlands Support to the Water Sector in India: Impacts on Institutional and Development and Change*, Haskoning, Netherlands.

Hazare, Anna (1997): *Ralegaon Siddhi: A Veritable Transformation translated by Pendse*, S.B. from original Marathi book, published by Ralegaon Siddhi Pariwar.

Heath, John (1989): "Privatization: Modalities and Strategy," in V.V. Ramanadham (ed.) *Privatization in Developing Countries*, Routledge, London, pp. 118-124.

Hirway, I (2005): "Ensuring Drinking Water to All: A Study in Gujarat." *4th IWMI-TATA Annual Partners Research Meet*, February 24-26.

— (1999), *Integrating Environmental & Economic Accounting: The Case of Gujarat – Phase 1*, Gujarat Ecology Commission, Vadodara, Prepared by Centre for Development Alternatives, Ahmedabad.

— and Shital Lodhia (2004): *Status of Drinking Water in Gujarat: Towards a Sustainable Approach*, Unpublished Report submitted to World Health Organisation, New Delhi.

— and P.P. Patel (1994): *Dynamics of Drinking Water in Rural Gujarat*, Action Aid India, Bangalore.

— and Subhrangsu Goswami (2008): "Functioning of the Drinking Water Component of the Narmada Pipeline Project in Gujarat," Special Article, *Economic and Political Weekly*, March 2008.

— *et al.* (2002): *Quick Valuation of Depletion and Degradation of Environmental Resources in Gujarat*, sponsored by Gujarat Ecology Commission, Vadodara, Prepared by Centre For Development Alternatives, Ahmedabad.

— and Subhrangsu Goswami (2005): "Performance of Narmada Pipeline Project in Gujarat", *Research Highlights No. 10*, International Water Management Institute, Anand.

— (2006 a): *Concurrent Monitoring of Worlds Largest Pipeline Drinking Water Project: Study of Narmada Pipeline Project in Gujarat*, Unpublished Draft Report submitted to International Water Management Institute, Anand and PRAVAH, Anand.

— (2006 b): *Status of Drinking Water in Coastal Saurashtra: Need for Area Specific Approach*, Unpublished Draft Report submitted to Coastal Salinity Cell, Rajkot.

— (2008): "Functioning of the Drinking Water Component of the Narmada Pipeline Project in Gujarat", Special Article, *Economic and Political Weekly*, March.

Hooja, Rakesh; Ganesh Pangare; and K.V. Raju (2002): *Users in Water Management: The Andhra Model and Its Replicability in India,* Rawat Publications. New Delhi, India.

Howe, Charles W, Dennis R. Schurmeier and W. Douglas Shaw, Jr. (1986) "Innovative Approaches to Water Allocation: The Potential for Water Markets," *Water Resources Research*, 22 (4), pp. 439-445, April.

Independent Review Mission (1991): *Independent Review, Sardar Sarovar.* World Bank.

Indian Express (2009): "Government orders removal of diesel pumps installed along Narmada canal.", Ahmedabad edition, June 20.

Indu, Rajnarayan and Alka Rawal (2007): "Impact of High Incidence of Kidney Stone: A Study of Coastal Villages in Junagadh, Gujarat", Paper presented in workshop on *Crisis in Drinking Water in Coastal India*, EDI, Ahmedabad.

Institute of Rural Management, Anand (IRMA) (2001): *White Paper on Water in Gujarat*, report prepared for Government of Gujarat, IRMA and UNICEF, Gandhinagar.

International Council of Scientific Unions (1972): *Man Made Lakes as Modified Ecosystem.* Scope Report 2. Paris.

International Water Power and Dam Construction (1988): *Reservior and the Environment—Experience in Managment and Monitoring.* Volume 40, Issue 9.

IPCC (Intergovernmental Panel on Climate Change) (2001): *Impacts, Adaptation, and Vulnerability*. Cambridge, England, Cambridge University Press.

IRMA – Institute of Rural Management (2001): *White Paper on Water in Gujarat*, prepared for the Department of Narmada, Water Resources and Water Supply, Government of Gujarat, Gandhinagar.

Iyengar, Sudarshan (1997): *Resettlement and Rehabilitation in SSP: A Review of Policy and Implementation.* GIDR. Ahmedabad.

—, and Vinayak Dave (1998): *Status of Project-affected Families from Madhya Pradesh Resettled in Gujarat.* Reconnaissance Report. GIDR. Ahmedabad.

Iyengar, Sudarshan (1998): *Gujarat* in V.M. Rao (ed.) *The Poor in a Hostile Society: Glimpses of Changing Poverty Scenario in India*, Vikas Publishing House Pvt. Ltd., New Delhi.

Jayaratne, R.S. (1990): "Involuntary Resettlement Consequent to the Construction of Three Large Dams and Reservoirs under the Mahaweli Development Project, Sri Lanka" in A.P. Fernandez (ed.) *Workshop on Rehabilitation of Persons Displaced by Development Projects: A Report*, Institute for Social and Economic Change and MYRADA. Bangalore.

Jodha, N.S. (1981): "Role of Credit in Farmers' Adjustment against Risk in Arid and Semiarid Tropical Areas of India," *Economic and Political Weekly*, Vol. 26 (42-43).

Johnson, III Sam H. (1997): "Irrigation Management Transfer: Decentrlizing Public Irrigation in Mexico," in *Water International*, 22(3), September, 159-67 and in Maria Saleth (Eds), 2002; *Water Resource and Economic Development*, Edward Elgar Publishing Ltd., UK.

Joshi, Archana (2006): *Groundwater Conservation: A Perspective Plan for Ahmedabad Urban Complex Area, Master of Planning* Dissertation Report (unpublished), School of Planning, CEPT University, Ahmedabad.

Joshi, M.B. (2001): "Operation of Sardar Sarovar Conveyance System", *International Journal of Water Resources Development*, Vol. 17, 1 March.

— and Kapadia, Vivek (2010): "Sharing Water in the 21st Century: Rethinking the Rationale", *Irrigation and Drainage*, 59: 92-101.

Joshi, Vidyut (1987): *Submerging Villages – Problems and Prospects*, Ajanta Publications (India). Delhi.

— (1991): *Rehabilitation – A Promise to Keep (A Case Study of SSP)*, The Tax Publications. Ahmedabad.

— (1991): *Rehabilitation at Sardar Sarovar Narmada Project* in Pathak, Mahesh T. (ed.): *Sardar Sarovar Project: A Promise for Plenty*. Oxford and IBH Publishing. New Delhi.

Kalpavriksh (1995): "Cry Our Beloved Narmada," *The India Magazine*, New Delhi. February.

Kay, Melvyn; Tom Franks and Laurence Smith (1997): *Water: Economics, Management and Demand*. London: E&FN Spon, London.

Khosla, A.N. (1964): Report submitted to the Central Water Commission. New Delhi.

Kothari, A. and S. Singh (1988): *The Narmada Valley Project: A Critique*, Kalpavriksh. New Delhi.

Kothari, Smitu (1996): "Whose Nation? Displaced as Victims of Development," *Economic and Political Weekly*, 31 (24).

Krishiworld.com http://www.krishiworld.com/html/water_resources2. html (accessed on 28th April, 2010)

Krishna Jai, Chaturvedr R.S., Srivastava L.S. (1976): *Geological and Seismological Investigations of the Environs of Narmada Valley around Navagam Dam Site in Gujarat State.*

—(1989): "Dams and Seismicity.".

Krishna Kumar (1996): "Narmada Debate: A Test of Modernity," *Indian Express*, July, 11.

Kumar, M. Dinesh (2002): Reconciling Water Use and Environment: Water Resource Management in Gujarat, Resource, Problems, Issues, Strategies and Framework for Action, *Report prepared for State Environmental Action Project* supported by the World Bank. Gujarat Ecology Commission.

— (2005): "Impact of Electricity Prices and Volumetric Water Allocation on Energy and Groundwater Demand Management: Analysis from Western India", *Energy Policy*, 33 (1).

— (2007): *Groundwater Management in India: Physical, Institutional and Policy Alternatives*, New Delhi: Sage Publications.

— (2009): *Water Management in India: What Works, What Doesn't*, New Delhi: Gyan Books.

— (2010): *Managing Water in River Basins: Hydrology, Economics, and Institutions*, New Delhi: Oxford University Press.

— and Jos van Dam (2009): "Improving Water Productivity in Agriculture in India: Beyond 'More Crop per Drop'", in Kumar, M. Dinesh and Upali Amarasinghe (Eds) Water Productivity Improvements in Indian Agriculture: Potentials, Constraints and Prospects, *Strategic Analysis of National River Linking Project of India Series 4*. Colombo: International Water Management Institute.

— and O.P. Singh (2001): "Market Instruments for Demand Management in the Face of Scarcity and Overuse of Water in Gujarat," *Water Policy*, 3 (5).

— (2005): "Virtual Water in Global Food and Water Policy Making: Is There a Need for Rethinking?" *Water Resources Management*, 19: 759-789.

— and Upali Amarasinghe (Eds) (2009): "Water Productivity Improvements in Indian Agriculture: Potentials, Constraints and Prospects", *Strategic Analysis of National River Linking Project of India Series 4*. Colombo: International Water Management Institute, Colombo.

— A. Narayanamoorthy, O.P. Singh, M.V.K. Sivamohan, Manoj Sharma and Nitin Bassi (2010), "Gujarat's Agricultural Growth Story: Exploding Some Myths", *Occasional Paper 2*, Institute for Resource Analysis and Policy, Hyderabad.

— Ajaya Kumar Malla and Sushanta Tripathy (2008): "Economic Value of Water in Agriculture: Comparative Analysis of a Water-Scarce and a Water-Rich Region in India", *Water International*, 33 (2): 214-230.

Kumar, M., A. Dinesh, A. Narayanamoorthy and M.V.K. Sivamohan (2010), Gujarat's *Agricultural Growth Story: Exploding Some Myths*. Occasional Paper 2, Institute for Resource Analysis and Policy, Hyderabad.

Kumar, M.D. (1995): "Dynamics of drinking water in rural Gujarat: physical, social and institutional dimensions", paper presented in the workshop of *Drinking Water Problems in Rural Gujarat: Issues and Perspectives*, GIDR, January 25.

Levenhagen, D.W. (1987): *Overview Environmental Impact Report: Sardar Sarovar and Narmada Sagar Complex Project Area, Narmada River Basin, India*. Consultant's Report. World Bank. Washington D.C.

Lobo, L., S. Kumar (2009): *Land Acquisition, Displacement and Resettlement in Gujarat,1947-2004,* SAGE Publications, New Delhi.

Lobo, Lancy and Shashikant Kumar (2009): *Land Acquisition, Displacement and Resettlement in Gujarat: 1947-2004*, Sage India, New Delhi.

Luders, Rolf J. (1991), "Massive Divestiture and Privatization: The Case of Chile," *Contemporary Policy Issues*, Vol. 9, No. 4, pp. 1-19.

M.S. University, Baroda (1993): *Studies on Ecology and Environment*, Department of Geology, Project Report submitted to SSNNL, Government of Gujarat.

Mahapatra, L.K. (1990): "Rehabilitation of Tribals Affected by Major Dams and Other Projects in Orissa," in A.P. Fernandez (ed.) *Workshop on Rehabilitation of Persons Displaced by Development Projects: A Report*, Institute for Social and Economic Change and MYRADA. Bangalore.

Mallik, Pradeep (2010): "Dams! Sarovar Project A Mess," *Ahmedabad Mirror*, Ahmedabad. March.

Manthan Adhyayan Kendra (2005): *Unravelling Bhakra*. (A Report), Badwani, Madhya Pradesh, India. April.

Marie-Helene, Z. (2006): *Urban Water and Waste Water*. New Delhi: Oxford University Press.

Mathew, A. (1989): "Industries, Mines and Disposition of Indigenous People," in W. Fernandes and Enakshi Ganguly Thukral (eds.) *Development, Displacement and Rehabilitation*, Indian Social Institute. New Delhi.

McKinsey Global Institute (2010): *India's Urban Awakening: Building Inclusive Cities, Sustaining Economic Growth*. April.

Meade, Ronald B. (1991): *Reservoirs and Earthquakes. Engineering Geology*. Volume 30, Issue 3-4.

Mehta, S.S. (2010): "Access to Urban Water and Sanitation to the Poor in Gujarat and Maharashtra". *A Research Report Prepared for PAS Project*, (draft, mimeo), CEPT University.

Meinzen-Dick Ruth S., L. Brown, H.S. Feldstein and A.R. Quisumbing (1997): "Gender, Property Rights and Natural Resources," in *World Development*, 25 (8), pp. 1303-1315.

Merrey, D.J. (1996): "Institutional Design Principles for Accountability in Large Irrigation Systems", *Research Report 8*. Colombo: International Water Management Institute.

MHUPA and UNDP (2009): *India: Urban Poverty Report 2009,* Oxford University Press, New Delhi.

Milgroms, Paul and Roberts, John (1992): *Economics, Organization & Management*, Prentice Hall Publication, pp 543.

Ministry of Agriculture and Cooperation (1985): *Report of the Inter-departmental Committee on Soil Conservation and Afforestation, Sardar Sarovar and Narmada Catchment.* Government of India. New Delhi.

Ministry of Environment & Forests (1986): *Environmental Aspects of Narmada Sagar and Sardar Sarovar* – A Study for Narmada Planning Group.

— (1986): *Cost Benefit Analysis.* Circular No.11-61/85-FRY (Cons). Government of India, January, 13.

— (1987a): Office Memorandum No. 3-87/80-1A. June, 24.

— (1987b): Memorandum No. 8-372/83-FC. September, 8.

— (1987c): Memorandum No 8-646//84-FC. October, 7.

— (1990): Memorandum No.8- 29/89-FC. March, 15.

Ministry of Irrigation and Power (1972): *Report of the Irrigation Commission.* Government of India. Volume III, Part I. New Delhi.

Ministry of Urban Development (2008): *Handbook of Service Level Benchmarks*, Government of India.

Mitra, Amit (1995): "Development Distress," *Down to Earth*, Vol. 3, No. 21.

Modi, N. (2011): *Convenient Action: Gujarat's Responses to Challenges of Climate Change*, Macmillan Publishers India, New Delhi.

Morse, Bradford and Thomas Berger (1992): *Sardar Sarovar: Report of the Independent Review*, Resource Futures International Inc. Ottawa.

Musgrave, Richard A. and Peggy B. Musgrave (1989): *Public Finance in Theory and Practice*, 5th Edition. McGraw-Hill Book Company. New York.

Muthayya, B.C. and J.S. Mathur (1975): *Rehabilitation of Displaced Villagers*, National Institute of Community Development. Hyerabad.

Nagi Reddy, C.P. *et al*. (1981): *Relocation and Resettlement Plan for Project Affected Persons Under Singoor Project*, Centre for Economic and Social Studies. Hyderabad.

Naik, P.J., R.S. Patil and R.S. Maradi (1990): "Reservoir Submergence and Rehabilitation in Six Villages in the Upper Krishna Project," in A.P. Fernandez (ed.) *Workshop on Rehabilitation of Persons Displaced by Development Projects: A Report*, Institute for Social and Economic Change and MYRADA. Bangalore.

Nankani, Helen B. (1990): "Lessons in Privatization in Developing Countries," *Finance and Development*, Vol. 27, No. 1, pp. 43-45.

Narmada Bachao Andolan (1990): *Narmada – A Campaign Newsletter*, New Delhi. December.

— (1992): "Response to ARCH-Vahini's Critique of Narmada Bachao Andolan and Anti-dam Arguments," *(Mimeo)*.

— (1990): *Sardar Sarovar Project: An Economic, Environmental and Human Disaster*. Bombay.

Narmada Control Authority (1991): *Minutes of Meeting: Review of Activities of the Environmental Subgroup*, Item No XXXVIII 6 (417).

— (1992): *Silting of Sardar Sarovar.* Publication Number 2/92

—(1993): *Seismicity and Sardar Sarovar Project.*

—(2007): *Status Report on Sardar Sarovar Project.*

Narmada Planning Group (1981): *Projectised Studies: Sardar Sarovar (Narmada) Project.* Government of Gujarat.

—(1989): *Planning for Prosperity: Sardar Sarovar Development Plan.* Gandhinagar.

—(1996): *Report of the Expert Committee on financial Resources for Sardar Sarovar Project.*

Narmada Control Board (1989): *Minutes of Meeting: Rehabilitation Subgroup*, Narmada Valley Development Authority. December.

Narmada Planning Group (n.d.): "Cropping pattern and Water demand study in Narmada Command Area", Volume-I *Base Studies*, Government of Gujarat, Gandhinagar.

Narmada Project Dam Design Circle (1982 and 1983): *Sedimentation Studies and Life of Reservoir: Sardar Sarovar (Narmada) Project (FRL 455 ft)*. Vadodara.

Narmada Valley Development Authority (1992): *Action Plan of Rehabilitation and Resettlement of Oustees of Madhya Pradesh (Sardar Sarovar Project)*, Parts I and II. Government of Madhya Pradesh. Bhopal. January.

— (1993): *The Bhil Track: A Study of Displaced Tribals*, Government of Madhya Pradesh. Bhopal.

Narmada Water Dispute Tribunal (1978): *Report of the Narmada Water Dispute Tribunal.* Government of India. New Delhi.

—(1978): *Further Report on the Narmada Water Dispute Tribunal with it's Decision.* Government of India. New Delhi.

Narmada Water Disputes Tribunal Award (1979): *Sub-Clause IV(1) – Provision for Rehabilitation*, December.

Narmada Water Disputes Tribunal. (1979): *The Report of the Narmada Water Disputes Tribunal,* Volumes 1 and 2. New Delhi.

National Commission on Farmers (2005): *Serving Farmers and Saving Farming – Jai Kisan: A Draft National Policy for Farmers*, Draft Fourth Report, Ministry of Agriculture, Government of India.

National Remote Sensing Centre (2009): *Assessment of Irrigation Potential created in ABIP funded Irrigation Projects in India using Cartosat Satellite data*, Hyederabad.

Nitin Desai Expert Committee (1983): *Report of the Committee to Review the Existing Criteria for Working out Benefit Cost Ratio for Irrigation Projects.* Government of India.

Padhiari, H., (2005): "Water Service Markets in Surface Irrigation Systems: Institutions and Socio-Economic Impacts." IWMI-Tata Water Policy Programme, paper presented in *3rd ITP Annual Partners' Meet*, (unpublished research) Anand.

Paranjpye, V. (1989): *The Narmada Valley Project: A Holistic Evaluation of the Sardar Sarovar and Indira (Narmada) Sagar Dams,* Indian National Trust for Art and Cultural Heritage (INTACH). New Delhi.

— (1990): *High Dams on the Narmada*, INTACH. New Delhi.

Paranjpye, Vijay (1990): *High Dams on the Narmada: A Holistic Analysis of the River Valley Projects*, Indian National Trust for Art and Cultural Heritage. New Delhi.

Parasuraman, S. (1993): "The Anti-Narmada Movement in India: Can the Resettlement and Rehabilitation Policy Gains be Translated into a national Policy," *Working Paper Series, No.161*, Institute of Social Studies, The Hague.

— (1996): "Outcomes for Displaced: Two Case Studies," *Economic and Political Weekly*, 31 (24).

Parthasarathy, R. (2000): "Participatory irrigation Management Programme in Gujarat: Institutional and Financial Issues," *Economic and Political Weekly*, Vol. 35 (35 and 36), August, 26 to September 2.

— (2002): "Reforms in Irrigation Management and Peoples' Participation," in Rakesh Hooja, Ganesh Pangare and K.V. Raju (ed.) *Users in Water Management: The Andhra Model and its Replicability in India,* Rawat Publications, Jaipur and New Delhi.

— (2008): "Irrigation Water Pricing: Analytic of Competing Sources," in Vishwa Ballabh (ed.) *Governance of Water: Institutional Alternatives and Political Economy,* Sage, New Delhi.

— and Jharna Pathak (2008a): "The Guiding Visible Hand of Participatory Approaches to Irrigation Management," in V Ratna Reddy and S Mahendra Dev (ed.) *Managing Water Resources: Policies, Institutions, and Technologies,* Oxford University Press, New Delhi.

— and P.K. Vishwanathan (2008b): "Are Water Policies A Case of Reverse Engineering in India?" *Proceedings of the 7th Annual Partners Meet Volume - 2,* IWMI-TATA Water Policy Research Programme, Hyderabad, April.

— (2010): "The Role of Irrigation in the Growth Story of Gujarat," in Ravindra Dholakia and Samar K. Datta (ed.) *High Growth Trajectory and Structural Changes in Gujarat Agriculture*, Macmillan Publishers, New Delhi.

Patel, A.S. and H.F. Patel (1992): *Some Policy Issues for Canal Water Irrigation in Gujarat.* Department of Economics, Sardar Patel University.

Patel, Anil (2001), "Resettlement in the Sardar Sarovar Project: A Cause Vitiated". *International Journal of Water Resources Development*, 17 (3): 315-328.

— *et al.* (1990): "The Tragic Story of the Displaced: A Case Study of the Tribals Affected by Sardar Sarovar," in A.P. Fernandez (ed.) *Workshop on Rehabilitation of Persons Displaced by Development Projects: A Report*, Institute for Social and Economic Change and MYRADA. Bangalore.

— (1991): "Rehabilitation and Resettlement in Sardar Sarovar Project: Are the Critics Right?" *Social Action.* New Delhi. July-September.

— (1994): "What do the Tribals Want in the Valley of Narmada?" Paper presented at an *International Conference on Indigenous People and Water Reservoir Development Projects*, Montreal, Canada, April.

— (1995): "Reforms or Derailment: Activists' Dilemma Revisited," Paper presented at Refugee Studies Programme (RSP) Conference on *Development induced Displacement and Improvement*, Oxford, U.K. January.

— (1997): "What do the Narmada Valley Tribals Want?" in W.F. Fisher (ed.) *Towards Sustainable Development*, Rawat Publications. Jaipur.

— (2001): "Resettlement in Sardar Sarovar Project," *Water Resource Development*, Vol. 7(3).

—; A. Mehta; S. Iyengar; T. Parekh and R. Mishra (1992): *Morse-Berger Review: A Search for Truth,* ARCH-Vahini. Mangrol.

Patel, H.G. (1990): *Financial Working of Irrigation Projects—A Case Study of Four Projects in Gujarat.* Department of Economics, Sardar Patel University.

Patel, V.M. (1991): "Sardar Sarovar Project : Benefit-Cost Analysis" in M.T. Pathak (ed.) *Sardar Sarovar Project: A Promise for Plenty*, Oxford and IBH Publishing Co. Pvt. Ltd., New Delhi.

Pathak, M., V.D. Shah (2010): "Five Decades of Gujarat Agriculture: Some Reflections" in R.K. Dholakia, S.K. Datta (Eds): *High Growth Trajectory and Structural Changes in Gujarat Agriculture*. Macmillan Publishers India Ltd.

Pathak, Mahesh (1989): *Environmental Planning for Sardar Sarovar project.* SSNNL. Gandhinagar.

—(ed.) (1991): *Sardar Sarovar Project: A Promise for Plenty.* Oxford and IBH Publishing. New Delhi.

—(ed.) (1995): *Development Imperatives and Environmental Concerns, Sardar Sarovar Project.* Agro Economic Research Centre. Vallabh Vidyanagar.

Patkar, Medha (1988): *Sardar Sarovar: Vikas Ki Vinash* (in Hindi), Parisar, Pune.

Perry, Chris J. (2001a): "World Commission on Dams: Implications for Food and Irrigation", *Irrigation and Drainage*, 50: 101-107.

— (2001b), Water at any Price? Issues and Options in Charging for Irrigation Water, *Irrigation and Drainage*, 50 (1): 1-7.

Plusquellec, H. (2002): *How design, management and policy affect the performance of irrigation projects: Emerging modernisation procedures and design standards*. Food and Agriculture Organisation of the United Nations, Bangkok ftp://ftp.fao.org/docrep/fao/004/ac799e/ac799e00.pdf (accessed on 3rd February, 2010) 20.

Raj, P. A. (1990): *Facts: Sardar Sarovar Project*, Sardar Sarovar Narmada Nigam, Gandhinagar.

Raj, P.A. (1991): Erroneous Zones, *All About Narmada*, Government of Gujarat, Gandhinagar.

Ram, Rahul (1993): *Muddy Waters,* Kalpavriksh. New Delhi.

Ranade, Rahul and M. Dinesh Kumar (2004), "Narmada Water for Groundwater Recharge in North Gujarat: Conjunctive Management in Large Irrigation Projects", *Economic and Political Weekly*, XXXIX (31): pp. 3498-3503.

Rao, S.V.L.N. (1992): *Pricing of a Public Utility: A Survey.* Department of Economics, South Gujarat University.

Raskin, P., P. Gleick, P. Kirshen, G. Pontius and K. Strzepek (1997): "Water Futures: Assessment of Long-Range Patterns and Problems", Stockholm Environment Institute, Stockholm.

Ravindra H. Dholakia and Samar K. Dutta (Eds) (2010): *High Growth Trajectory and Structural Changes in Gujarat Agriculture*, Macmillan Publishers, New Delhi.

Reddy, I.U.B. (1992): *Displacement and Rehabilitation*, Mittal Publications. New Delhi.

— (1994): *Industrial Development and Problems of the Uprooted*, Rawat Publications. Jaipur.

Rich, Bruce (1988): *Statement before the Subcommittee on Foreign Operations Committee on Appropriations*, US Senate. June, 14.

Rondinelli, Dennis A. and John D. Kasarda (1991): Privatizing Public Services in Developing Countries: What Do We Know?", *Business & the Contemporary World*, Vol. 3, No. 2, Winter, pp. 102-113

Rosegrant, Mark W. and Hans P. Binswanger (1994): "Markets in Tradable Water Rights: Potential for Efficiency Gains in Developing Country Water Resource Allocation", *World Development*, 22 (11): pp. 1613-1625.

— and Renato Gazmuri Schleyer (1994), "Reforming Water Allocation Policy through Markets in Tradable Water Rights: Lessons from Chile, Mexico, and California", EPTD Discussion paper 2, International Food Policy Research Institute, Washington.

— (1996) Establishing Tradable Water Rights: Implementation of the Mexican Water Law, *Irrigation and Drainage Systems*, 00: 263-279.

RTI India (2008): Gujarat used only 7% Narmada dam water. As reported in Yahoo News 29th May, 2008. http://www.rtiindia.org/forum/4391-gujarat-used-only-7-narmada-dam-water.html (accessed on 3rd January, 2010).

Sah, D.C. (1993): *Monitoring and Evaluation of Resettlement and Rehabilitation Programme for Sardar Sarovar (Narmada) Project*, Report 16. Centre for Social Studies. Surat.

— (1994a): *Monitoring and Evaluation of Resettlement and Rehabilitation Programme for Sardar Sarovar (Narmada) Project*, Report 19. Centre for Social Studies. Surat.

— (1994b): "Avoiding Impoverishment Risks: Towards a Policy Frame of Dam Affected People," Paper presented for the Seminar on *Development, Displacement and Rehabilitation* at Centre for Social Studies. Surat. 11-13 May. Revised version of the paper was accepted for publication in *Review of Development and Change* [Vol. 3 (1) 1998].

— (1994c): *Commercialisation in Agriculture and Living Condition,* (M&E Report 21), Centre for Social Studies. Surat.

— (1995a): *Monitoring and Evaluation of Resettlement and Rehabilitation Programme for Sardar Sarovar (Narmada) Project*, Report 20. Centre for Social Studies. Surat.

— (1995b): "Development and Displacement: National Rehabilitation Policy." *Economic and Political Weekly*, Vol. 30 (48). December 2.

— (1996a): *Monitoring and Evaluation of Resettlement and Rehabilitation Programme for Sardar Sarovar (Narmada) Project*, Report 21. Centre for Social Studies. Surat.

— (1996b): *Monitoring and Evaluation of Resettlement and Rehabilitation Programme for Sardar Sarovar (Narmada) Project*, Draft Report 23. Centre for Social Studies. Surat.

— (1996c): "Resettlement and Rehabilitation: Sardar Sarovar Project on the Narmada" *(Mimeo)*, Report No. 22. Centre for Social Studies. Surat.

— (1996d): *Living Condition in 31 New Sites* (M & E Report 24), Centre for Social Studies. Surat.

— (1997): "Second Generation Problems of Involuntary Migration," *Indian Journal of Agricultural Economic,* Vol. 52(4).

— (1998): "Avoiding Impoverishment Risks: Towards a Policy Frame of Dam Affected People," *Review of Development and Change*, Vol. 3(1).

— (1999): "Pressures on Land, Employment and Migrations," *Review of Development and Change,* Vol. 4(2).

— (2003): *Involuntary Migration: Evidence from Sardar Sarovar Project,* Rawat Publications, Jaipur.

— and J.N. Soni (1993): *Monitoring and Evaluation of Resettlement and Rehabilitation Programme for Sardar Sarovar (Narmada) Project*, Report 15. Centre for Social Studies. Surat.

Saleth, R.M. (1999): "Irrigation Privatization in India: Options, Framework and Experience", *Economic and Political Weekly*, 34(26): A86-A92.

Sardar Sarovar Narmada Nigam Limited (SSNNL) (1988): *Planning for Prosperity. Sardar Sarovar Development Plan*, Narmada Planning Group, Gandhinagar.

Scudder, Thayer (1991): "A Sociological Framework for the Analysis of New Land Settlements" in Michael M. Cernea (ed.) *Putting People First: Sociological Variables in Rural Development*, Oxford University Press, Second Edition. New York.

— (2003): *India's Sardar Sarovar Project,* (Unpublished Manuscript).

Sen, A.K. (1981): *Poverty and Famines,* Oxford University Press. Oxford.

Shah, G. and Biswaroop Das (1986): "Voluntary Organisation Development," *(Mimeo)*. Centre for Social Studies. Surat.

— (1988): "Voluntary Organisation in Gujarat," *(Mimeo)*. Centre for Social Studies. Surat.

Shah, H. (2009): *Narmada Yojana: Nehro ne badle pipeline ketli vyajabi?* (Gujarati) Arthbodh.

Shah, T. (1993): "Groundwater Markets and Irrigation Development", *Political Economy and Practical Policy*. Oxford University Press, Bombay.

— (2009): *Taming the Anarchy. Groundwater Governance in South Asia*, Washington DC: RFF Press.

—, S. Bhatt, R.K. Shah, J. Talati (2008): "Groundwater Governance through Electricity Supply Management: Assessing an Innovative Intervention in Gujarat, Western India". *Agriculture Water Managegment,* 95(11), 1233-1242.

—, S. Bhattacharya (1993): *Farmer Organisations for Lift Irrigation: Irrigation Companies and Tubewell Cooperatives of Gujarat*, Network Paper 26. Overseas Development Institute, London.

—, S. Verma (2008): "Co-management of Electricity and Groundwater: An Assessment of Gujarat's Jyotirgram Scheme". *Economic and Political Weekly,* Vol.43(7), 59-66.

Shah, T., A. Gulati, P. Hemant, G. Shreedhar, R.C. Jain (2009): "Secret of Gujarat's Agrarian Miracle after 2000". Review of Agriculture. *Economic and Political Weekly*, Vol. XlIV, no. 52.

Shah, Tushaar (1991): "Water Markets and Irrigation Development in India," *Indian Journal of Agricultural Economics*, Vol. 46, No. 3, July-September.

— (1993): "*Ground Water Markets and Irrigation Development in India: Political Economy and Practical Policy,"* Oxford University Press. Bombay.

Shah, V.I., Desai, N.V. and S.V. Joshi (1997): *Control of Brackishness in Drinking Water*, Central Salt and Marine Chemicals Research Institute (CSMCRI), Bhavnagar.

Shah, Zankhana and M. Dinesh Kumar (2008), "In the Midst of the Large Dam Controversy: Objectives, Criteria for Assessing Large Water Storages in the Developing World", *Water Resources Management*, 22 (12): 1799-1824.

Sharan, Girja (2007): *Harvesting Dew to Supplement Drinking Water Supply in Arid Coastal Villages of Gujarat, Paper presented in workshop on "Crisis in Drinking Water in Coastal India*, EDI, Ahmedabad.

Sharma, U.S. (1996): "Some Aspects of Drinking Water Supply in Rural Gujarat (Adequacy and Operation – Maintenance)", Proceedings of National Seminar On *Rural Water Supply Sanitation*, Organised by Centre For Development Studies, Thiruvananthapuram.

Sharma, Usha (1995): *A Study on Some Aspects of Rural Drinking Water Supply in Gujarat: Adequacy, Quality and Operation-Maintenance*, Sardar Patel Institute of Economic and Social Research, Ahmedabad.

Sheth, Pravin (1991): *Sardar Sarovar Project: Dynamics of Development*, Vikas Bharati Institute of Policy Studies, Research and Futurology. Ahmedabad.

— (1995): "Politics of Narmada Project," *(Mimeo)*. Madhya Pradesh Institute of Social Science Research. Ujjain.

Shingi, Prakash M. and V.N. Asopa (2002): *Independent Evaluation of Check dams in Gujarat Strategies and Impacts*, Centre for Management in Agriculture, Indian Institute of Management, Ahmedabad.

Shroff, Manu (1995): *Economic Viability of SSP* in Pathak, Mahesh (ed.) *Development Imperatives and Environmental Concerns*. Agro Economic Research Centre. Vallabh Vidyanagar.

Shunglu, V.K., G.K. Chadha and Jaiprakash Narayan (2006): *Report on the Status of Rehabilitation of Project Affected Families in Madhya Pradesh*, Sardar Sarovar Project Relief & Rehabilitation Oversight Group.

Singh, O.P. (2004): "Water-intensity of North Gujarat's Dairy Industry: Why Dairy Industry Should Take a Serious Look at Irrigation?", Paper presented at the Second International Conference of the *Asia Pacific Association of Hydrology and Water Resources* (APHW 2004), Singapore, June 5-9.

Singh S, (2003): "Large Dams and Environment" in Prasad K (ed.) *Water Resource and Sustainable Development*, Shipra Publications, Delhi.

— (2006): "Credit, Indebtedness and Farmer Suicides in Punjab: Some Missing Links." *Economic and Political Weekly*. 41(3), 3330–31.

Singhal, N., V. Patwari, V. (2009): "Evolving arrangements for local water diversion-delivery in SSP." *Management Traineeship Segment Report*, Institute of Rural Management, Anand.

Sinha, B.K. (1996): "National Policy for Rehabilitation: Objectives and Principles," *Economic and Political Weekly*, 31 (24).

Sneh Setu Foundation (2008): *Impact of Sardar Sarovar Project Water on Selected Villages of Gujarat*, A study report, September, Vadodara.

Solanki, Rupa (1994): *Condition of Tribal Women in Relocation Sites*, A report submitted towards partial fulfilment of coursework in Master of Social Work, Gujarat Vidyapith. Ahmedabad.

Soni, Trupti and Gautam Appa. (1993): "Sardar Sarovar Project Downstream Impact," Paper presented in *Workshop at Narmada Forum*, Delhi School of Economics and Institute of Economic Growth.

SSNL (1988): *Planning for Prosperity. Sardar Sarovar Development Plan*, Narmada Planning Group, Gandhinagar.

— (2003): *Environment & Amelioration Programme (Fisheries)*, Gandhinagar, January. No. 1," in *Texas Canal Lining and Pipeline Installation Project*", http://www.cocef.org/aproyectos/ExComHidalgo2003_05ing.htm (accessed on 28th December, 2009)

— (2009): "Benefits-Irrigation", Sardar Sarovar Narmada Nigam Limited. <http://www.sardarsarovardam.org/benefits.htm> (accessed on 12 October, 2009).

— (2009): *Water Use Plan—SSP*. Gandhinagar.

—(2010): *Report of the Expert Group for Strategy for Accelerated Development of Sardar Sarovar Project command Area*. Gandhinagar.

— (2010): Personal discussions with Executive Director, Agriculture, Sardar Sarovar Narmada Nigam Limited, Gandhinagar.

SSNNL's letter of 24/04/98 to Executive Member, NCA.

Sudarsen, V. and M.A. Kalam (eds.) (1990): *The Uprooted: Development, Displacement and Resettlement,* Gian Publishing House. Delhi.

Supplementary Affidavit to the Writ Petition (Civil) 319. 319 (Supreme Court of India, July 1997).

Supreme Court of India (2002): *Arundhati Roy.... Contemner, Contempt Petition (crl.) 10 of 2001*, Judgment, March 6, 2002, New Delhi.

SWRC (1988): "Towards People's Power in Jhabua," *People's Action*. May.

TAHAL Consulting Engineers Limited, Israel (1987): *Water Resources Planning for the State of Gujarat,* A report submitted to SSNNL, Gandhinagar, Gujarat.

Talati, J., J. Liebrand (2003): "Evolving institutions for irrigation management in Sardar Sarovar Project Command", *IWMI-Tata Water Policy Program*,(unpublished research) Anand.

—, T. Shah (2004): "Institutional Vacuum in Sardar Sarovar Project: Framing 'Rules-of-the-Game'". *Economic and Political Weekly*, 39(31), 3504–509.

Tata Economic Consultancy Services (1981): *Methodological Framework for Economic Appraisal of Narmada Project.*

Tata Institute of Social Sciences (1993): *The Sardar Sarovar Project: Experiences with Resettlement and Rehabilitation*, Tata Institute of Social Sciences. Bombay.

Thakkar, Himanshu (2005): "Who Pays for Our Giant Follies?" *Info-change News & Features,* October.

The Hindu (2006): *GoM's confidential report on A Brief Note on the Assessment of Resettlement and Rehabilitation (R&R) Sites and Submergence of Villages of the Sardar Sarovar Project.* Note of the Group of Ministers Dated 9 April 2006. Chennai. April 17.

Thobani, Mateen (1997), "Formal Water Markets: Why, When and How to Introduce Tradable Water Rights", *The World Bank Research Observer*, 12 (2): 161-179.

Thukral, Enakshi Ganguly (1989): "Dams for Whose Development?" in W. Fernandes and E. Ganguly Thukral (eds.) *Development, Displacement and Rehabilitation: Issues for a National Debate*, Indian Social Institute. New Delhi.

— (ed.) (1992): *Big Dams, Displaced People: Rivers of Sorrow Rivers of Change,* Sage Publications. New Delhi.

Times of India (2009a): "Shelve Narmada pipeline plan: Expert." Ahmedabad edition, September 12.

— (2009b): "Narmada pipeline project scaled down." Ahmedabad edition, November 13.

Tolba, M.K. (1988): *Sustainable Water Development: Opportunities and Constraints*, Vol. 13 (4).

TRENDSNIFF. (2008, September): Retrieved December 2010, from www.trendsniff.com: http://trendsniff.com/2008/09/16/targeting-top-20-growth-cities-in-india/

United States Department of the Interior, Bureau of Reclamation (1976): *Design of Gravity Dams.* Denver.

— (1983): *Economic Appraisal of Sardar Sarovar Project.* Bombay.

Unknown (2009): "State to Press for a Taller Narmada Dam," *Indian Express*. October 23.

US Bureau of Reclamation (1973): *Water System Automation*, Current Information on Automation of Reclamation Water Systems, United States Department of the Interior, Denver, Colorado.

Vaidyanathan, A. (2006): *India's Water Resources – Contemporary Issues on Irrigation*, Oxford University Press, New Delhi.

Varghese, B.G. (1994): *Winning the Future: From Bhakra to Narmada, Tehri, Rajasthan Canal*, Konark Publishers Pvt. Ltd., New Delhi.

Varma, S.C. (1985): *Human Settlement in Lower Narmada Basin*, Narmada Valley Development Authority. Bhopal.

Vermillion, Douglas (1997): *Impacts of Irrigation Management Transfer*, IIMI, Research Report 11, April.

Vice Chairman, NVDA's letter No. NVDA/D (Fish)/88/3867 of 1/11/88 addressed to the Chairman, NCA.

Vyas, J. (2001), "Water and energy for development in Gujarat with special focus on the Sardar Sarovar project", *International Journal Water Resources Development*, 17(1):37–54.

Vyas, Jay Narayan (2001): "Water and Energy for Development in Gujarat with Special Focus on the Sardar Sarovar Project," *International Journal of Water Resources Development,* Vol. 17, 1, March.

Wallingford, H.R. (1993): *Environmental Changes Downstream of Sardar Sarovar Dam*. Ex 2750.

—(1993): *Sardar Sarovar Project Hydrology and Water Resources.*

—(1993): *Sediment and Backwater Aspects of SSP Report.* Ex 2761.

—(1993): *SSP: Command Area Environmental Impact Assessment Progress Report.*

—(1995): *Potential for Agro Chemical Pollution from the Sardar Sarovar Command Area, India.* Ex 3177.

Wallingford, H.R. (1995): *Environmental Changes Downstream of Sardar Sarovar*. Oxfordshire, UK. Report No. EX 2750. of World Bank.

Wamana Consultants (1994): *Study of Preparation of Detailed Command Area Development Plan: SSP.*

2030 Water Resources Group (2009): *Charting Our Water Future: Economic Frameworks to Inform Decision-Making*, 2030WaterResourcesGroup @mckinsey.com

Watt, Jacqueline (2008), "The Effect of Irrigation on Surface-Ground Water Interactions: Quantifying Time Dependent Spatial Dynamics in Irrigation Systems," Thesis submitted to Charles Sturt University for the Degree of Doctor of Philosophy, School of Environmental Sciences, Faculty of Sciences, Charles Sturt University, March.

Wood, John R. (1993): "India's Narmada River Dam: Sardar Sarovar under Siege," *Asian Survey*, Vol. 33, No. 10.

World Bank (1985): *Narmada River Development, Gujarat—Sardar Sarovar Dam and Power Project*, Staff Appraisal Report 5107-IN; and *Water Delivery and Drainage Project*, Staff Appraisal Report 5108-IN.

—(1990): *Review of Sardar Sarovar (Narmada) Projects Economic Analysis*. India Agriculture Operations Division.

—(1993): *Project Completion Report: Narmada River Development, Gujarat.*

—(1995): *Project Completion Report-India-Narmada River Development-Gujarat Sardar Sarovar Dam and Power Project*, Report No. 14159.

— (1995): *Project Completion Report (Sardar Sarovar).* World Bank.

— (2005): *India's Water Economy: Bracing for a Turbulent Future*, World Bank, Washington.

Zoology Department (1994): *Environmental Impact Assessment Studies on inland and Marine Fisheries Relevant to the Command Area of Sardar Sarovar Project.* M.S. University. Baroda.

Index